D0212754

Mathematical Chestnuts from Around the World

Library of Congress Catalog Card Number 00-108053

Complete Set ISBN 0-88385-300-0
Vol. 24 ISBN 0-88385-330-2

Printed in the United States of America

Current printing (last digit):
10 9 8 7 6 5 4 3 2 1

The Dolciani Mathematical Expositions

NUMBER TWENTY-FOUR

Mathematical Chestnuts from Around the World

Ross Honsberger
University of Waterloo

Published and distributed by
The Mathematical Association of America

THE
DOLCIANI MATHEMATICAL EXPOSITIONS

Published by
THE MATHEMATICAL ASSOCIATION OF AMERICA

The DOLCIANI MATHEMATICAL EXPOSITIONS series of the Mathematical Association of America was established through a generous gift to the Association from Mary P. Dolciani, Professor of Mathematics at Hunter College of the City University of New York. In making the gift, Professor Dolciani, herself an exceptionally talented and successful expositor of mathematics, had the purpose of furthering the ideal of excellence in mathematical exposition.

The Association, for its part, was delighted to accept the gracious gesture initiating the revolving fund for this series from one who has served the Association with distinction, both as a member of the Committee on Publications and as a member of the Board of Governors. It was with genuine pleasure that the Board chose to name the series in her honor.

The books in the series are selected for their lucid expository style and stimulating mathematical content. Typically, they contain an ample supply of exercises, many with accompanying solutions. They are intended to be sufficiently elementary for the undergraduate and even the mathematically inclined high-school student to understand and enjoy, but also to be interesting and sometimes challenging to the more advanced mathematician.

---

1. *Mathematical Gems,* Ross Honsberger
2. *Mathematical Gems II,* Ross Honsberger
3. *Mathematical Morsels,* Ross Honsberger
4. *Mathematical Plums,* Ross Honsberger (ed.)
5. *Great Moments in Mathematics (Before 1650),* Howard Eves
6. *Maxima and Minima without Calculus,* Ivan Niven
7. *Great Moments in Mathematics (After 1650),* Howard Eves
8. *Map Coloring, Polyhedra, and the Four-Color Problem,* David Barnette
9. *Mathematical Gems III,* Ross Honsberger
10. *More Mathematical Morsels,* Ross Honsberger
11. *Old and New Unsolved Problems in Plane Geometry and Number Theory,* Victor Klee and Stan Wagon
12. *Problems for Mathematicians, Young and Old,* Paul R. Halmos
13. *Excursions in Calculus: An Interplay of the Continuous and the Discrete,* Robert M. Young
14. *The Wohascum County Problem Book,* George T. Gilbert, Mark Krusemeyer, and Loren C. Larson
15. *Lion Hunting and Other Mathematical Pursuits: A Collection of Mathematics, Verse, and Stories by Ralph P. Boas, Jr.,* edited by Gerald L. Alexanderson and Dale H. Mugler
16. *Linear Algebra Problem Book,* Paul R. Halmos
17. *From Erdős to Kiev: Problems of Olympiad Caliber,* Ross Honsberger
18. *Which Way Did the Bicycle Go? . . . and Other Intriguing Mathematical Mysteries,* Joseph D. E. Konhauser, Dan Velleman, and Stan Wagon
19. *In Pólya's Footsteps: Miscellaneous Problems and Essays,* Ross Honsberger
20. *Diophantus and Diophantine Equations,* I. G. Bashmakova (Updated by Joseph Silverman and translated by Abe Shenitzer)

21. *Logic as Algebra,* Paul Halmos and Steven Givant
22. *Euler: The Master of Us All,* William Dunham
23. *The Beginnings and Evolution of Algebra,* I. G. Bashmakova and G. S. Smirnova (Trans. by Abe Shenitzer)
24. *Mathematical Chestnuts from Around the World,* Ross Honsberger

MAA Service Center
P. O. Box 91112
Washington, DC 20090-1112
1-800-331-1622 fax: 301-206-9789

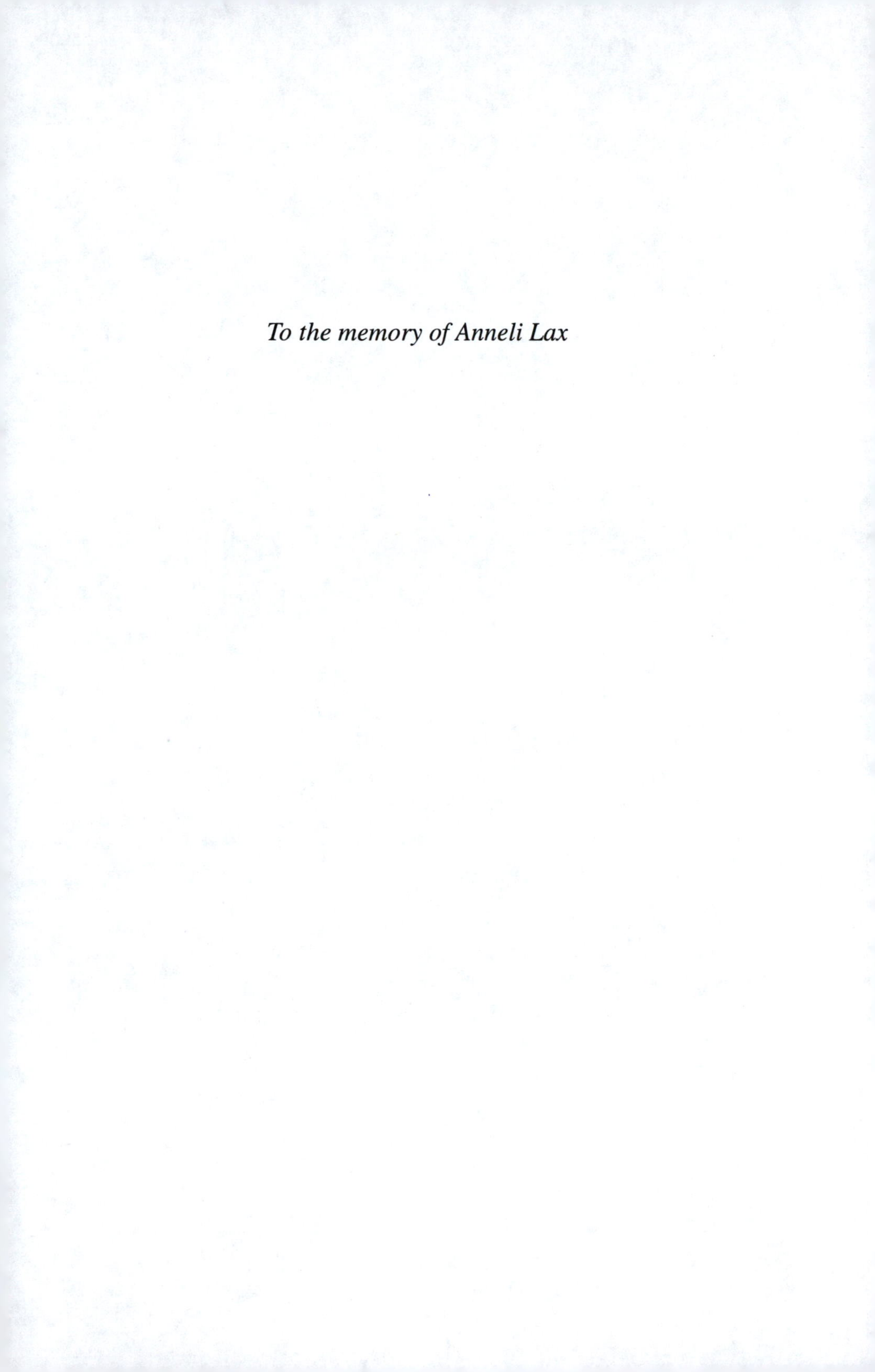

To the memory of Anneli Lax

Contents

Preface

This miscellaneous collection of elementary gems contains brilliant insights from many fine mathematical minds. Its more than 150 topics come from Euclidean geometry, combinatorics and combinatorial geometry, algebra and number theory, and most of the discussions can be followed comfortably by a college freshman.

There is no attempt to give instruction; in the few places where preliminaries are presented it is done only in preparation for a gem to follow.

The essays, written in a leisurely style, are intended as mathematical entertainment. While a measure of concentration is the price of enjoying some of these explanations, it is hoped that they will combine the excitement and richness of elementary mathematics with reading pleasure. Although it is not necessary to try the problems before going on to the solutions, if you are able to give them a little thought first, I'm sure you will find them all the more exciting.

The sections are independent and may be read in any order.

The problems are not grouped according to subect or arranged in a special order. There is a subject-index at the end to help you locate a particular interest.

It is a pleasure to thank the members of the Dolciani Editorial Board for their warm reception and gentle criticism of the manuscript. Again my sincerest thanks go to Elaine Pedreira and Beverly Ruedi for their unfailing geniality and expertise; what a difference they make!

SECTION 1
Five Problems from Ireland

The problems in this essay appeared on tests that were used in the selection of the 1990 Irish International Olympiad Team. It is a pleasure to thank Professor Finbarr Holland of University College, Cork, for his generosity in providing several of these solutions and for supplying the credits.

1. (Proposed by Tom Laffey, University College, Dublin.)

   As the name implies, the sequence of non-squares is determined from the positive integers by omitting the perfect squares:

   $$2, 3, 5, 6, 7, 8, 10, 11, 12, 13, 14, 15, 17, 18, \ldots.$$

   Prove that a formula for the nth non-square t_n is $[n + \frac{1}{2} + \sqrt{n}]$, where $[x]$ denotes the greatest integer $\leq x$.

Obviously a non-square lies strictly between two consecutive squares. Suppose

$$m^2 < t_n < (m+1)^2.$$

That is to say, the first t_n positive integers consist of m perfect squares and n non-squares. Thus

$$t_n = m + n.$$

Now, for k an integer, the fractional part of $(k + x)$ must be contained in the number x, and hence

$$[k + x] = k + [x].$$

Therefore

$$\left[n + \frac{1}{2} + \sqrt{n}\right] = n + \left[\frac{1}{2} + \sqrt{n}\right],$$

and it remains to show that

$$m = \left[\frac{1}{2} + \sqrt{n}\right].$$

This is equivalent to

$$m \leq \frac{1}{2} + \sqrt{n} < m + 1,$$
$$m - \frac{1}{2} \leq \sqrt{n} < m + \frac{1}{2},$$
$$m^2 - m + \frac{1}{4} \leq n < m^2 + m + \frac{1}{4},$$

and since m and n are integers, to

$$m^2 - m + 1 \leq n \leq m^2 + m.$$

But this follows immediately from the bounds on t_n:

$$m^2 < t_n < (m+1)^2.$$

Thus

$$m^2 < m + n < m^2 + 2m + 1,$$
$$m^2 - m < n < m^2 + m + 1,$$

and

$$m^2 - m + 1 \leq n \leq m^2 + m.$$

2. (This is an exercise from R. C. J. Nixon's *Euclid Revisited*, Oxford University Press, 1899, page 423; it was discovered there by Fergus Gaines, University College, Dublin, who also supplied the following beautiful solution.)

   Three sides of a convex quadrilateral $ABCD$ have lengths $AB = a$, $BC = b$, and $CD = c$. If the area of the quadrilateral is as large as possible, prove that the length x of the fourth side satisfies the equation

$$x^3 - (a^2 + b^2 + c^2)x - 2abc = 0.$$

When $ABCD$ has maximum area, suppose the length of $AC = t$. Now, in Figure 1,

$$\Delta ACD = \frac{1}{2}tc \sin \angle ACD \leq \frac{1}{2}tc \sin 90^\circ = \Delta ACD'.$$

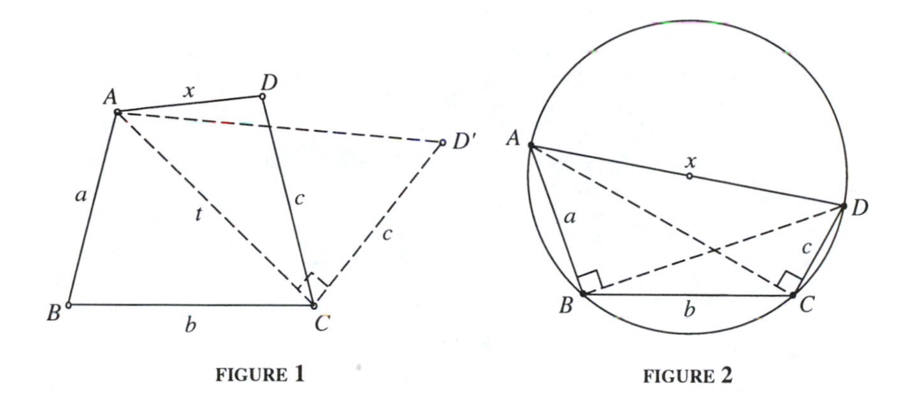

FIGURE 1

FIGURE 2

Therefore, if $\angle ACD$ is not a right angle, $\triangle ACD < \triangle ACD'$ and $ABCD'$ would be a quadrilateral with the prescribed dimensions which has an area greater than $ABCD$. Hence for maximum area, $\angle ACD$ must be a right angle; that is to say, the fourth side x must subtend a right angle at C.

Similarly, x must subtend a right angle at B, and it follows that for maximum area, $ABCD$ must be inscribed in a circle of diameter $AD = x$ (Figure 2).

Now, by the theorem of Ptolemy we have

$$AC \cdot BD = AB \cdot CD + BC \cdot AD.$$

From right triangles ACD and ABD, then, this yields

$$\begin{aligned}
\sqrt{x^2 - c^2} \cdot \sqrt{x^2 - a^2} &= ac + bx, \\
(x^2 - c^2)(x^2 - a^2) &= (ac + bx)^2, \\
x^4 - (a^2 + c^2)x^2 + a^2c^2 &= a^2c^2 + 2abcx + b^2x^2, \\
x^4 - (a^2 + b^2 + c^2)x^2 - 2abcx &= 0,
\end{aligned}$$

and since $x \neq 0$, we obtain the desired

$$x^3 - (a^2 + b^2 + c^2)x - 2abc = 0.$$

3. (Proposed by Gordon Lessells, Limerick University.)

$\triangle ABC$ is right angled at A and AD is the altitude from A (Figure 3). If E is the midpoint of DC and AB is extended its own length to F, prove that FD and AE are perpendicular.

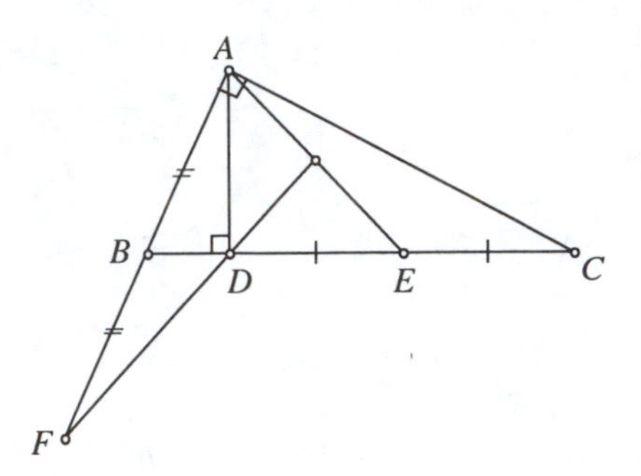

FIGURE 3

With the midpoints of AF and DC figuring prominently, one's suspicion might be aroused that other midpoints could be important. Accordingly, if M is the midpoint of AD (Figure 4), then ME joins the midpoints of two sides of $\triangle ADC$ and is therefore parallel to the third side AC. But AC is given perpendicular to AB, and so EM is also perpendicular to AB, in which case it lies on the altitude from E in $\triangle ABE$. But in $\triangle ABE$, AD is the altitude from A, and so M is actually the orthocenter of $\triangle ABE$.

Thus BM lies along the third altitude of $\triangle ABE$, making BM perpendicular to AE. But because B and M are midpoints of sides of $\triangle AFD$, BM is parallel to FD, and it follows that FD is also perpendicular to AE.

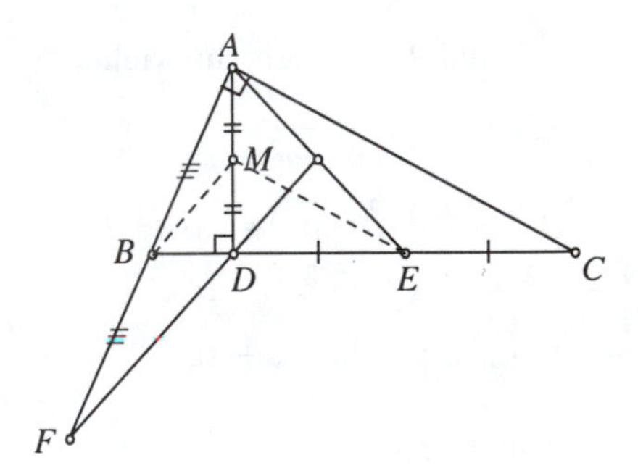

FIGURE 4

4. (Proposal and solution by Tom Laffey, University College, Dublin.)

For all positive integers $n \geq 3$, prove that

$$S_n = \frac{1}{3^3} + \frac{1}{4^3} + \frac{1}{5^3} + \cdots + \frac{1}{n^3} < \frac{1}{12}.$$

Since

$$(k-1)(k+1) = k^2 - 1 < k^2,$$

then

$$(k-1)k(k+1) < k^3,$$

and

$$\frac{1}{(k-1)k(k+1)} > \frac{1}{k^3}.$$

Therefore

$$S_n < \frac{1}{2 \cdot 3 \cdot 4} + \frac{1}{3 \cdot 4 \cdot 5} + \cdots + \frac{1}{(n-1)n(n+1)} = \sum_{k=3}^{n} \frac{1}{(k-1)k(k+1)}.$$

At this point one is tempted to expand $\frac{1}{(k-1)k(k+1)}$ into the partial fractions $\frac{A}{k-1} + \frac{B}{k} + \frac{C}{k-1}$ in the hope of obtaining a telescoping series. This is certainly a valid approach. However, Professor Laffey greatly simplifies the solution with the clever device of using the form $\frac{A}{(k-1)k} + \frac{B}{k(k+1)}$ for the partial fractions.

Accordingly, suppose

$$\frac{1}{(k-1)k(k+1)} = \frac{A}{(k-1)k} + \frac{B}{k(k+1)}.$$

In this case,

$$1 = A(k+1) + B(k-1),$$

and $k = -1$ and $+1$ provide the equations

$$1 = -2B \quad \text{and} \quad 1 = 2A, \quad \text{giving} \quad A = \frac{1}{2} \quad \text{and} \quad B = -\frac{1}{2}.$$

Thus

$$S_n < \frac{1}{2} \sum_{k=3}^{n} \left[\frac{1}{(k-1)k} - \frac{1}{k(k+1)} \right], \qquad \text{(a telescoping series)}$$

and

$$\begin{aligned} 2S_n &< \left(\frac{1}{2 \cdot 3} - \frac{1}{3 \cdot 4} \right) + \left(\frac{1}{3 \cdot 4} - \frac{1}{4 \cdot 5} \right) + \cdots + \left[\frac{1}{(n-1)n} - \frac{1}{n(n+1)} \right] \\ &= \frac{1}{2 \cdot 3} - \frac{1}{n(n+1)} \\ &< \frac{1}{6}, \end{aligned}$$

yielding the desired $S_n < \frac{1}{12}$.

5. (Proposed by Gordon Lessells and Jim Leahy, Limerick University.)

   Suppose the perpendiculars from C to the bisectors of angles A and B in $\triangle ABC$ meet them at D and E (Figure 5). Prove that DE is parallel to AB.

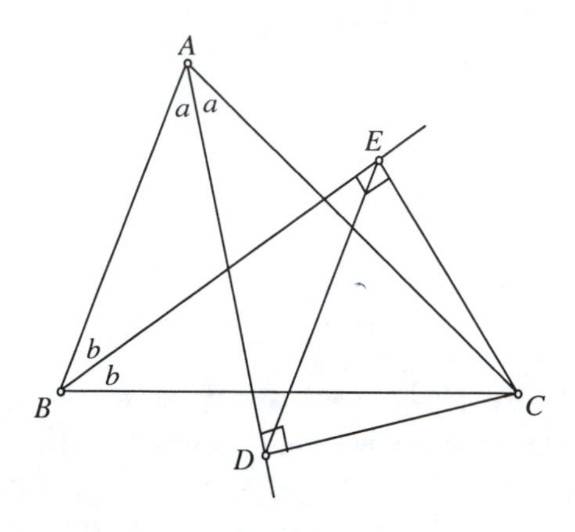

FIGURE 5

The bisectors of angles A and B meet at the incenter I of $\triangle ABC$, implying that IC is the bisector of $\angle C$ (Figure 6). Let $\angle A = 2a$, $\angle B = 2b$, and $\angle C = 2c$; then $2a + 2b + 2c = 180°$ and $a + b + c = 90°$. Also, let $\angle IDE = x$. Now, the right angles at D and E make $IDCE$ cyclic, and so

$$x = \angle IDE = \angle ICE \qquad \text{(in the same segment).}$$

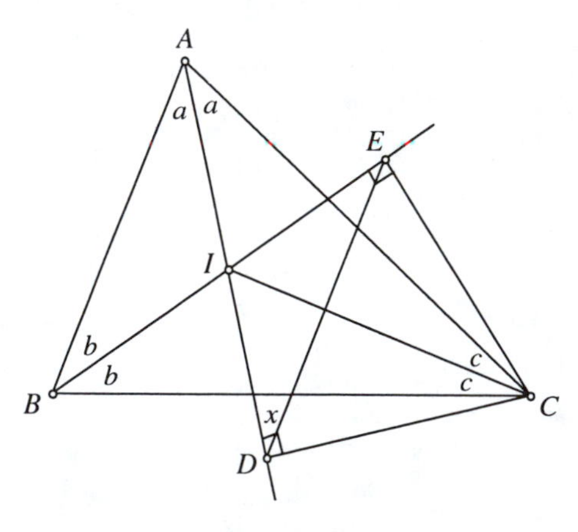

FIGURE 6

But, in right triangle BEC,

$$\angle BCE = 90^\circ - b.$$

Therefore

$$\begin{aligned}\angle ICE &= \angle BCE - c \\ &= 90^\circ - b - c \\ &= a \qquad \text{(recall } a + b + c = 90^\circ\text{)},\end{aligned}$$

and we have

$$x = a,$$

from which it follows that DE and AB are parallel (alternate angles at A and D).

SECTION 2

Three Solutions to a Variation on an Old Chestnut

The following problem, which is deceptively difficult, has been going the rounds for decades.

> In $\triangle ABC$, $AB = AC$ and $\angle A = 20°$ (Figure 1). A $60°$-angle is drawn at B and a $50°$-angle at C to give points D and E on AC and AB. Determine the size of angle $x = \angle EDB$.

This problem is the subject of Constantine Knop's delightful article "Nine Solutions To One Problem" that appeared in the May–June, 1994, issue of *Quantum*. At the end of the article we are given the following variant as an exercise.

> In the same isosceles triangle ABC, the base BC is laid off along AC from A to make $AP = BC$ (Figure 2). Determine the angle $\alpha = \angle PBC$.

This lovely problem was proposed by the ninth grade student Sergey Yurin.

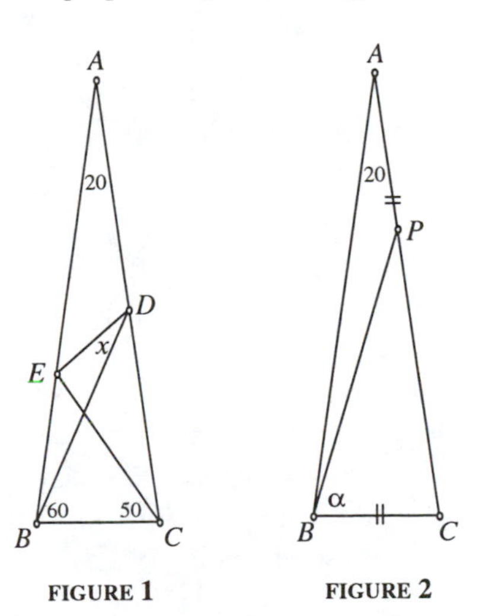

FIGURE 1 FIGURE 2

Our first solution is due to the ingenious Vladimir Dubrovsky of Moscow State University; it appeared in the May-June 1994 issue of *Quantum*.

1. Dubrovsky's Solution

First, observe that the base angles of $\triangle ABC$ are each $\frac{1}{2}(180° - 20°) = 80°$. Now, on the circle with center A and radius AB, let BC be stepped off twice from C to give points D and E (Figure 3); this reproduces $\triangle ABC$ as triangles ACD and ADE. Since $AB = AE$ and the three 20° angles at A add up to 60°, triangle ABE is equilateral, and we have $BE = AB$. Also, $AP = BC = DE$, giving a second pair of equal sides in triangles APB and BED. But it is easy to see that their included angles are also equal, making the triangles congruent:

> BD subtends 40° at the center A, and therefore 20° at E on the circumference, giving $\angle BED = 20° = \angle BAP$.

Thus $\angle ABP = \angle DBE$, and since DE subtends a 20° angle at the center A, we have $\angle DBE = 10°$ at B on the circumference. Finally, then, $\angle ABP = 10°$, and

$$\alpha = \angle PBC = \angle ABC - \angle ABP = 80° - 10° = 70°.$$

This is certainly a beautiful solution. Now let's turn to another brilliant solution, due to my colleague Ian McGee.

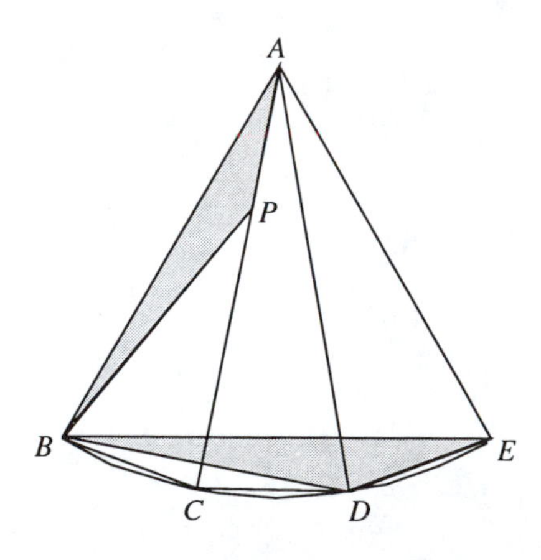

FIGURE 3

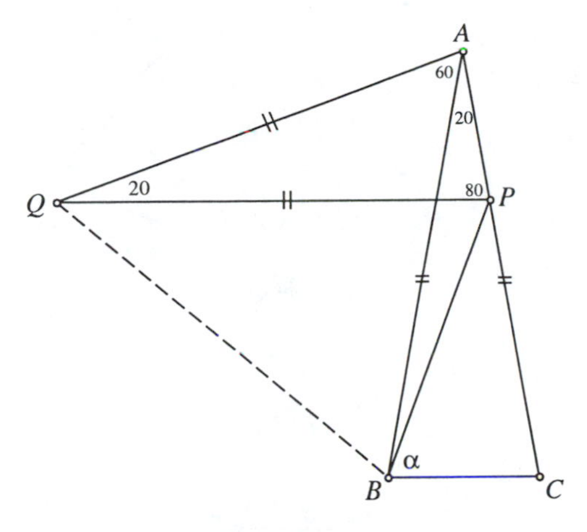

FIGURE 4

2. McGee's Solution

Again we begin with the observation that the base angles in $\triangle ABC$ are 80°. Now, since AP is equal to the base BC, Ian picks up $\triangle ABC$ and puts it down with BC on top of PA; this carries A to some point Q (Figure 4).

With base angle $QAP = 80°$ and $\angle BAC = 20°$, then $\angle QAB = 60°$. Hence in $\triangle AQB$, the angle between the equal sides AQ and AB is 60°; thus the triangle is equilateral and QB is the same length as QA and QP.

It follows, then, that Q is the center of a circle through A, P, and B, and since AP subtends a 20° angle at the center Q, it subtends a 10° angle at B on the circumference. Thus

$$\alpha = 80° - 10° = 70°.$$

3. Solution 3

The cue for our final solution comes from the remarkable approach taken by S. T. Thompson (Tacoma, Washington) in his 1951 solution of the original old chestnut (reported in my *Mathematical Gems II*, Dolciani Series, 1976, 16–18).

In the circle with center A and radius AB, let BC be stepped off 18 times around the circumference to give a wheel of 18 copies of $\triangle ABC$ ($18 \cdot 20° = 360°$) (Figure 5).

Now, it certainly appears that BP extended passes through the vertex X of the 8th copy of $\triangle ABC$ (counting $\triangle ABC$ itself). Of course this needs to be proved, but in the event it turns out to be true, we could argue as follows:

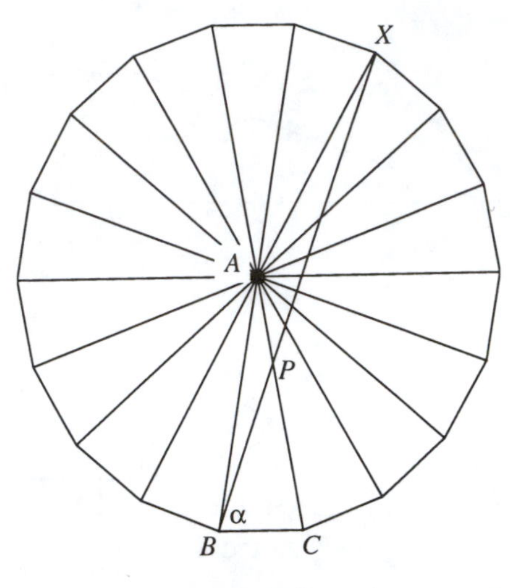

FIGURE 5

Clearly

$$\angle BAX = 8 \cdot 20^\circ = 160^\circ,$$

making the base angles at B and X in isosceles triangle BAX equal to 10°, implying $\alpha = 80^\circ - 10^\circ = 70^\circ$.

Instead of trying to prove that BP goes through X, let's join B and X and show that BX crosses AC at a point L such that $AL = BC$, thus identifying L and P.

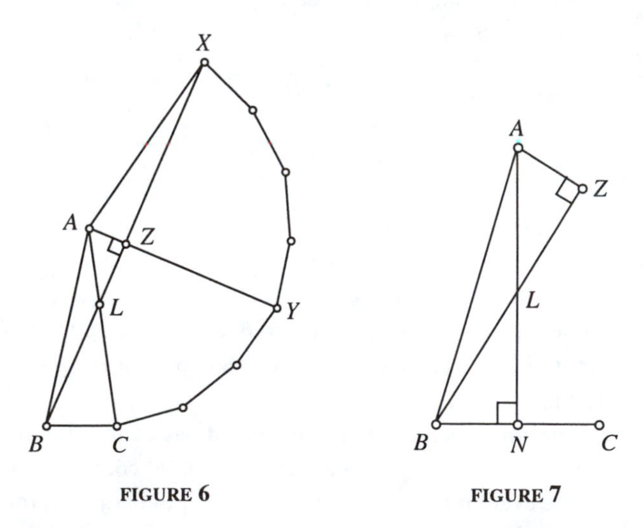

FIGURE 6

FIGURE 7

In Figure 6, there are four copies of our triangle between AB and AY, making $\angle BAZ = 4 \cdot 20° = 80°$. Since $\angle ABX = 10°$ (noted above), then $\angle AZB$ is a right angle. But clearly, $\angle LAZ = \angle CAY = 3 \cdot 20° = 60°$, and so $\triangle LAZ$ is a 30°-60°-90° triangle with $AL = 2AZ$. It remains to show, then, that BC is equal to $2AZ$.

To this end, let AN be the altitude to base BC in our isosceles triangle ABC (Figure 7). Then N bisects BC, and we need to show that $BN = AZ$. But this is immediate, for triangles ABZ and ABN are congruent:

each triangle has angles of 80° and 90°, and side AB is common.

SECTION 3

Three Problems from Eötvös-Kürschák Competitions

1. (This problem, from Hungary's 1953 Eötvös-Kürschák Competition, is taken with solution from the article "Some Problems Memorable To Me" by János Surányi, which appeared in the first issue in 1993 of the *Journal of the World Federation of National Mathematics Competitions*.)

   If n is a positive integer and d is a divisor of $2n^2$, prove that $n^2 + d$ cannot be a perfect square.

Since d is a divisor of $2n^2$, then, for some positive integer k, $2n^2 = dk$. Now, if $n^2 + d$ were to be a perfect square, so would $k^2(n^2 + d)$ be a perfect square (a simple observation, but really very clever). However,

$$\begin{aligned} k^2(n^2 + d) &= k^2n^2 + k^2d \\ &= k^2n^2 + k(2n^2) \\ &= n^2(k^2 + 2k) \\ &= n^2[(k+1)^2 - 1], \end{aligned}$$

which would require $[(k+1)^2 - 1]$ to be a perfect square. But no two consecutive positive integers are both perfect squares, and since $(k+1)^2$ **is** a square, the $[(k+1)^2 - 1]$ can't be, and the conclusion follows by contradiction.

2. (From the 1898 Competition)

   Four points, A, B, C, D, are chosen at random on a straight line L. Construct a square $PQRS$ such that A and B, respectively, lie on the sides PQ and RS (possibly extended) and similarly C and D lie on QR and PS (Figure 1).

Extending a pair of opposite sides of a rectangle determines a parallel strip across the plane. A square thus determines two such strips which are the same

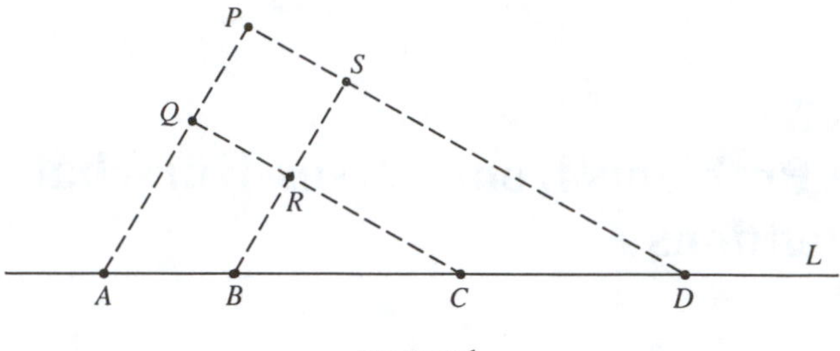

FIGURE 1

size. This problem, then, requires two equal perpendicular strips, with edges respectively through the pairs (A, B) and (C, D). Thus one might be inclined to give some thought to the way in which the width of a strip is measured.

We are so accustomed to using the perpendicular distance QR from one edge to the other that we are seldom inclined to consider other possibilities. However, the width is given equally well by the length of any transversal AB, provided the angle θ it makes with an edge is specified (Figure 2); certainly two strips with equal transversals at the same angle are the same size.

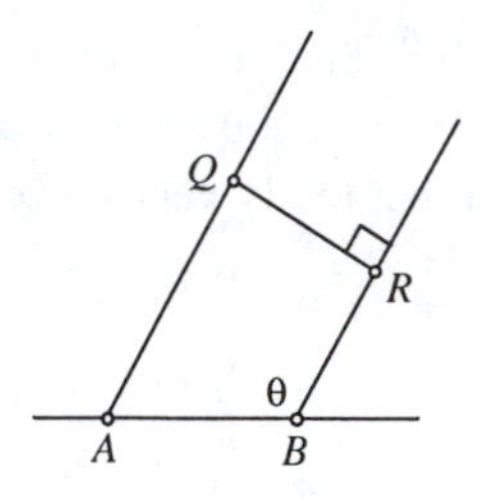

FIGURE 2

Now, let a transversal CX, perpendicular to L, be drawn across the strip determined by PS and QR (Figure 3). Then the right angles at C and S imply

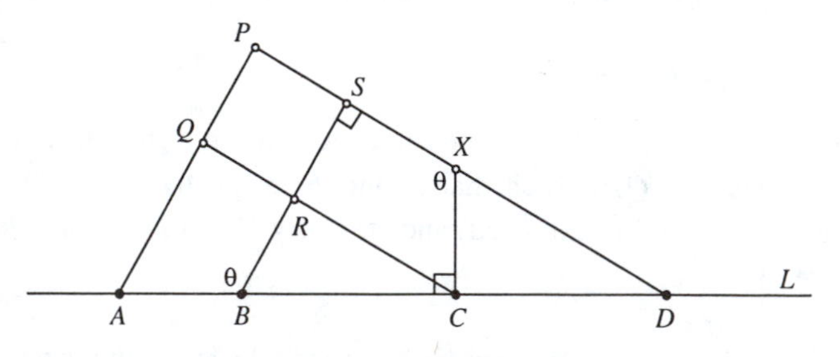

FIGURE 3

$BCXS$ is a cyclic quadrilateral, giving equal angles θ at B and X. Thus CX and AB are transversals that cross the equal strips at the same angle, and it follows that $CX = AB$.

Hence it is only necessary to make CX the same length as AB, after which the construction is completed by joining DX and drawing a parallel through C and perpendiculars through A and B.

3. (Now for a beautiful problem from the 1908 Competition; our solution is taken from the *Hungarian Problem Book II*, which was assembled by G. Hájos, G. Neukomm, and J. Surányi, translated by Elvira Rapaport and published as Volume 12 of the New Mathematical Library Series of the MAA. It is discussed here with the permission of the Mathematical Association of America.)

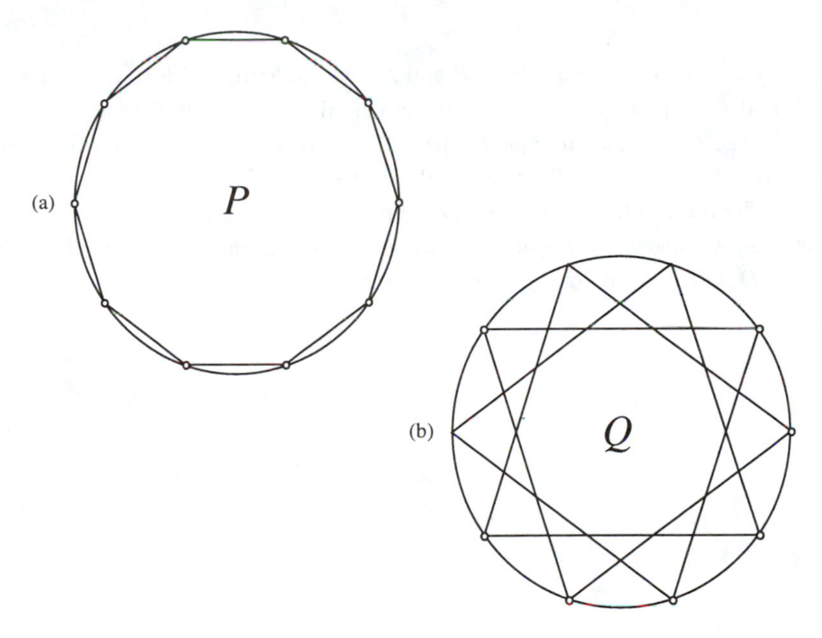

FIGURE 4

If 10 equally-spaced points around a circle are joined consecutively, a convex regular inscribed decagon P is obtained (Figure 4a); if every third point is joined, a self-intersecting regular decagon Q is formed (Figure 4b). Prove that the difference between the length of a side of Q and the length of a side of P is equal to the radius of the circle.

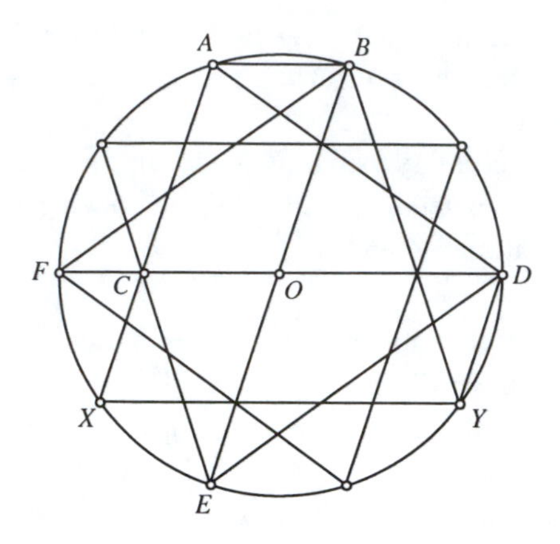

FIGURE 5

In Figure 5, AB is a side of P and XY is a side of Q. Clearly the chords BE and FD divide the points equally and are diameters of the circle.

Since the points are equally-spaced around the circle, clearly the three chords AB, FD, and XY are parallel. Similarly, AX, BE, and DY are parallel. Therefore $ABOC$ and $CXYD$ are parallelograms, and so the difference between AB and XY is simply the difference between their opposite sides CO and CD, which is obviously the radius OD.

SECTION 4

Three Problems from the Polish Mathematical Olympiads of 1949–1954

These problems appeared on early Polish olympiads, 1949–1954, and their solutions are based on the methods given in the wonderful book *Mathematical Problems and Puzzles from the Polish Mathematical Olympiads* by S. Straszewicz, published in 1965 by Pergamon Press as Volume 12 in their series of Popular Lectures in Mathematics. This series can also boast of a volume by each of the great mathematicians Waclaw Sierpinski and Hugo Steinhaus, and it is absolutely first-class in every way. I've never seen a book with more comprehensive treatment of its topics than this volume by Dr. Straszewicz.

1. If A, B, C, and D are four consecutive vertices of a regular polygon such that

$$\frac{1}{AB} = \frac{1}{AC} + \frac{1}{AD},$$

how many sides does the polygon have?

First, observe that if the polygon had only the four vertices A, B, C, D, then it would be a square and the given relation would fail, for then AB and AD would be equal, requiring the impossible $\frac{1}{AC} = 0$. Hence the polygon has at least one more vertex E following D, and since the polygon is regular, it is cyclic (Figure 1).

Let x be the length of a side of the polygon, $y = AC$, and $z = AD$. Then

$$\frac{1}{AB} = \frac{1}{AC} + \frac{1}{AD}$$

yields

$$\frac{1}{x} = \frac{1}{y} + \frac{1}{z}$$

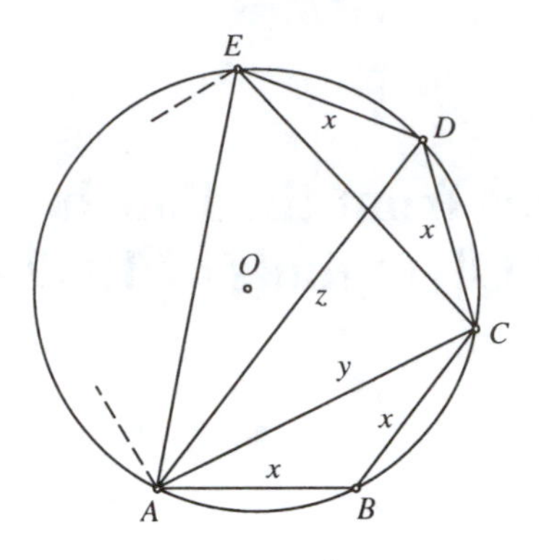

FIGURE 1

and we have

$$yz = xz + xy. \qquad (1)$$

Now, applying the theorem of Ptolemy to quadrilateral $ACDE$, we get

$$AE \cdot x + xy = EC \cdot z.$$

However, by symmetry, we have $EC = AC = y$, and so

$$\begin{aligned} AE \cdot x + xy &= yz \\ &= xz + xy \qquad \text{(by 1)} \end{aligned}$$

giving

$$AE \cdot x = xz$$

and

$$AE = z = AD.$$

Thus A is equidistant from D and E, in which case A lies on the perpendicular bisector of DE. But the perpendicular bisector of a chord goes through the center O of the circle. Therefore, going around the circle from $A \to B \to \cdots$ to the other end of the diameter through A, that is, halfway around, to the midpoint of DE, we count $3\frac{1}{2}$ sides of the polygon. Thus altogether the polygon has $2 \cdot 3\frac{1}{2} = 7$ sides.

2. If a plane figure has exactly two axes of symmetry, prove they must be perpendicular.

Let P be any point of a plane figure F which has axes of symmetry XY and UV which are **not** perpendicular. Clearly these axes are either parallel or they intersect, and since these are essentially equivalent cases, let us consider only the case in which XY and UV intersect (at O; Figure 2).

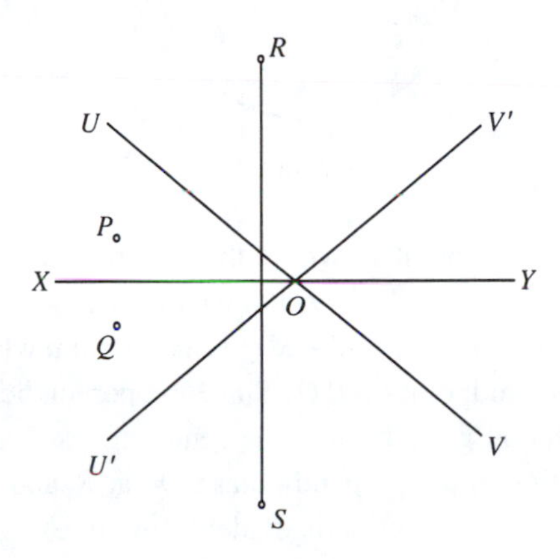

FIGURE 2

Let Q be the image of P in XY, R the image of Q in UV, and S the image of R in XY; then P, Q, R, and S are all points of F. Now, reflection in XY takes QR to PS and at the same time carries the perpendicular bisector UV of QR into the perpendicular bisector $U'V'$ of PS. That is to say, reflection in the line $U'V'$ would take P to an image point S in the given figure F. Thus a pair of non-perpendicular axes of symmetry implies a third axis ($U'V'$) and therefore the number of axes could be limited to two only if they are perpendicular (in which case $U'V'$ and UV are the same line (Figure 3)).

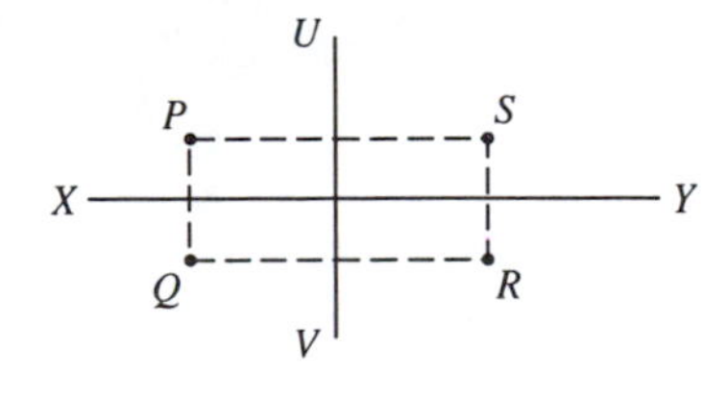

FIGURE 3

3. A and B are points on opposite sides of a straight line m. Construct a circle through A and B to intercept m in a chord PQ of minimum length (Figure 4).

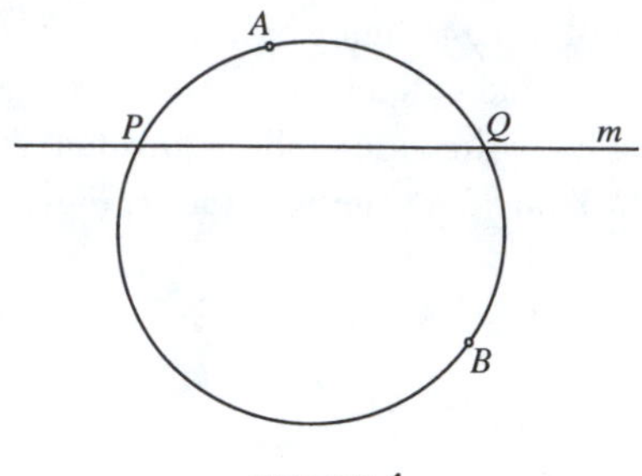

FIGURE 4

It is always the same point K where the variable chord PQ crosses AB (Figure 5). Thus the product $PK \cdot KQ$ of the parts of PQ always has the same value $AK \cdot KB$. Hence their sum $PK + KQ$ is a minimum when they are equal, that is, when K is the midpoint of PQ. Since the perpendicular to a chord of a circle from its midpoint goes through the center, the center O of the desired circle is the intersection of the perpendicular to m at K and the perpendicular bisector of AB, and the construction is evident (Figure 6).

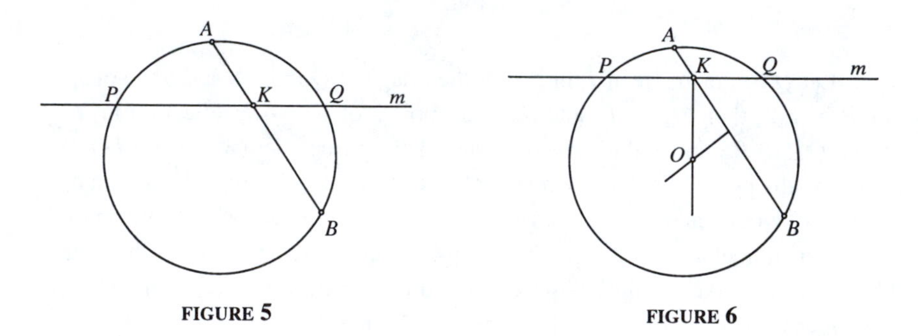

FIGURE 5

FIGURE 6

SECTION 5
Ten Problems from East German Olympiads

The problems in this section come from the eightieth volume of *Junge Mathematiker*, whose cover bills it as the one hundred most beautiful problems from the East German Olympiads for grades 11 and 12. The collection, published in German in 1987, was compiled by five former contestants of these olympiads and it was seen through publication by J. Lehmann of Leipzig. Many of the problems and solutions are indeed beautiful; the following is just a sample of its treasures. I am indebted to Professor Dr. H.-D. Gronau of Rostock, himself one of the compilers, for giving me a copy of this intriguing volume.

All but one of these solutions come from *Junge Mathematiker*. As usual, I have taken the liberty of writing them up to suit myself.

1. Consider the sum

$$N = 1 + \frac{1}{2} + \frac{1}{3} + \frac{1}{4}.$$

If one clears of fractions by multiplying through by 4!, we get

$$\begin{aligned} N \cdot 4! &= \left(1 + \frac{1}{2} + \frac{1}{3} + \frac{1}{4}\right) \cdot 24 \\ &= \frac{12 + 6 + 4 + 3}{12} \cdot 24 = \frac{25}{12} \cdot 24 = 50, \end{aligned}$$

a number which is divisible by 5 $(= 4 + 1)$.

Prove that, for odd integers $n \geq 3$,

$$N(n-1)! = \left(1 + \frac{1}{2} + \frac{1}{3} + \cdots + \frac{1}{n-1}\right)(n-1)!$$

is always divisible by n.

Since n is odd, the $n-1$ terms in N go together into $\frac{n-1}{2}$ pairs to yield

$$N = \left(1 + \frac{1}{n-1}\right) + \left(\frac{1}{2} + \frac{1}{n-2}\right) + \left(\frac{1}{3} + \frac{1}{n-3}\right) + \cdots$$
$$+ \left[\frac{1}{(n-1)/2} + \frac{1}{(n+1)/2}\right]$$
$$= \frac{n}{1(n-1)} + \frac{n}{2(n-2)} + \frac{n}{3(n-3)} + \cdots + \frac{n}{[(n-1)/2][(n+1)/2]}.$$

Since the product of all the denominators here is just $(n-1)!$, clearing of fractions at this point yields terms which have a common factor n, and the conclusion follows.

2. Suppose $f(x)$ is a function defined on the closed unit interval $[0, 1]$ such that

$$f(x) \geq 0, \quad f(1) = 1,$$

and, for all x_1, x_2 in $[0, 1]$ for which $x_1 + x_2$ is also in $[0, 1]$,

$$f(x_1 + x_2) \geq f(x_1) + f(x_2).$$

Prove that

$$f(x) < 2x \quad \text{for} \quad 0 < x \leq 1.$$

Since $x_2 \geq 0$ and $f(x_2) \geq 0$, then from the given

$$f(x_1 + x_2) \geq f(x_1) + f(x_2),$$

we have that

$$f(x_1 + x_2) \geq f(x_1),$$

implying that $f(x)$ is **nondecreasing** as x increases.

Since $f(1) = 1$, we have

$$f\left(\frac{1}{2^0}\right) = \frac{1}{2^0}.$$

Now, if $f(\frac{1}{2^k}) \leq \frac{1}{2^k}$ for some $k \geq 0$, then

$$\frac{1}{2^k} \geq f\left(\frac{1}{2^k}\right) = f\left(\frac{1}{2^{k+1}} + \frac{1}{2^{k+1}}\right)$$
$$\geq f\left(\frac{1}{2^{k+1}}\right) + f\left(\frac{1}{2^{k+1}}\right) = 2 \cdot f\left(\frac{1}{2^{k+1}}\right),$$

implying

$$f\left(\frac{1}{2^{k+1}}\right) \leq \frac{1}{2^{k+1}}.$$

Therefore, by induction, we have

$$f\left(\frac{1}{2^k}\right) \leq \frac{1}{2^k} \quad \text{for all} \quad k = 0, 1, 2, \ldots,$$

guaranteeing the relation $f(x) < 2x$ for $x = \frac{1}{2^k}$.

For all other x in the domain we have, for some $k \geq 0$, that

$$\frac{1}{2^{k+1}} < x < \frac{1}{2^k}.$$

Therefore, if $f(x) \geq 2x$ for any such x, we would have

$$f(x) \geq 2x > 2 \cdot \frac{1}{2^{k+1}} = \frac{1}{2^k} \geq f\left(\frac{1}{2^k}\right),$$

that is,

$$f(x) > f\left(\frac{1}{2^k}\right), \quad \text{where } x < \frac{1}{2^k},$$

contradicting the nondecreasing character of $f(x)$. Hence $f(x) \geq 2x$ fails for all $x \in (0, 1]$.

3. For a fixed positive integer $n \geq 2$, prove that the function

$$\begin{aligned} f(x) &= \sum_{k=1}^{n} \cos(x\sqrt{k}) \\ &= \cos(x\sqrt{1}) + \cos(x\sqrt{2}) + \cdots + \cos(x\sqrt{n}) \end{aligned}$$

is **not** periodic for real values of x.

Proceeding indirectly, suppose $f(x)$ is periodic. In this case, for some positive real number p, we have

$$f(x + p) = f(x) \qquad \text{for all values of } x.$$

For $x = 0$, this yields

$$f(p) = f(0) = \sum_{k=1}^{n} \cos(0) = \sum_{k=1}^{n} 1 = n;$$

that is to say,

$$f(p) = \cos(p\sqrt{1}) + \cos(p\sqrt{2}) + \cdots + \cos(p\sqrt{n}) = n.$$

Since $\cos\theta$ is never greater than 1, this could only be true if each term $\cos(p\sqrt{k}) = 1$. In particular, we must have both

$$\cos(p\sqrt{1}) = 1 \quad \text{and} \quad \cos(p\sqrt{2}) = 1.$$

But $\cos\theta = 1$ only for $\theta = 2\pi s$, an integral multiple of 2π. Thus, for some integers s and t,

$$p\sqrt{1} = 2\pi s \quad \text{and} \quad p\sqrt{2} = 2\pi t.$$

Dividing these results we get the contradiction that

$$\sqrt{2} = \frac{t}{s}, \qquad \text{a **rational** number.}$$

Thus $f(x)$ cannot be periodic.

4. Determine whether the **number** of positive integral solutions (a, b, c, d, e) of the equation

$$a^3 + b^5 + c^7 + d^{11} = e^{13}$$

is zero, a finite positive number, or infinite.

Sometimes equations like this have solutions in which all the terms on the expanded side are equal. While it is not clear how to accomplish

$$a^3 = b^5 = c^7 = d^{11},$$

if it can be done, the left side would amount to $4a^3$, and that would require e to have a factor 2. In an attempt to keep things as simple as possible, perhaps we can manage to have each term on the left a power of 2. Thus we might try for

$$(2^s)^3 = (2^t)^5 = (2^u)^7 = (2^v)^{11},$$

which requires

$$3s = 5t = 7u = 11v.$$

In this case, s would have to be divisible by 5, 7, and 11, making it at least $5 \cdot 7 \cdot 11$, and $a \geq 2^{5 \cdot 7 \cdot 11}$; similarly for t, u, and v. Thus, let's check

$$a = 2^{5 \cdot 7 \cdot 11}, \quad b = 2^{3 \cdot 7 \cdot 11}, \quad c = 2^{3 \cdot 5 \cdot 11}, \quad \text{and } d = 2^{3 \cdot 5 \cdot 7}.$$

This gives

$$a^3 = b^5 = c^7 = d^{11} = 2^{3 \cdot 5 \cdot 7 \cdot 11},$$

and the left side adds up to

$$4 \cdot 2^{3 \cdot 5 \cdot 7 \cdot 11} = 2^{3 \cdot 5 \cdot 7 \cdot 11 + 2} = 2^{1157} = (2^{89})^{13},$$

giving at least one solution

$$(a, b, c, d, e) = (2^{385}, 2^{231}, 2^{156}, 2^{105}, 2^{89}).$$

Now, if (a, b, c, d, e) satisfy

$$a^3 + b^5 + c^7 + d^{11} = e^{13},$$

then, multiplying through by $n^{3 \cdot 5 \cdot 7 \cdot 11 \cdot 13}$, n any positive integer > 1, we get

$$(a \cdot n^{5 \cdot 7 \cdot 11 \cdot 13})^3 + (b \cdot n^{3 \cdot 7 \cdot 11 \cdot 13})^5 + (c \cdot n^{3 \cdot 5 \cdot 11 \cdot 13})^7 + (d \cdot n^{3 \cdot 5 \cdot 7 \cdot 13})^{11} = (e \cdot n^{3 \cdot 5 \cdot 7 \cdot 11})^{13},$$

giving another solution

$$(an^{5005}, bn^{3003}, cn^{2145}, dn^{1365}, en^{1155}).$$

Thus, perhaps surprisingly, the number of solutions is infinite!

5. Find all functions $f(x)$, defined on the real numbers, such that
   (i) $f(1) = 1$,
   (ii) $f(x_1 + x_2) = f(x_1) + f(x_2)$,
   and, for $x \neq 0$,
   (iii) $f\left(\frac{1}{x}\right) = \frac{1}{x^2} f(x)$.

First of all, we have

$$f(0) = f(0 + 0) = f(0) + f(0) = 2f(0),$$

giving

$$f(0) = 0.$$

In this case,

$$0 = f(0) = f(x - x) = f[x + (-x)] = f(x) + f(-x),$$

implying

$$f(-x) = -f(x).$$

Now, property (iii) yields $f(x) = x^2 f(\frac{1}{x})$, which asserts that the value of the function at any point $x \neq 0$ is given by the product of the square of the argument x and the value of the function at the point $\frac{1}{x}$; thus, for example, if x is also not equal to 1, we have

$$f\left(\frac{1-x}{x}\right) = \left(\frac{1-x}{x}\right)^2 f\left(\frac{x}{1-x}\right).$$

Hence, for all $x \neq 0$ or 1, we have the following remarkable chain of results:

$$\begin{aligned}
f(x) &= x^2 f\left(\frac{1}{x}\right) \\
&= x^2 f\left(\frac{x+1-x}{x}\right) \\
&= x^2 f\left(1 + \frac{1-x}{x}\right) \\
&= x^2\left[f(1) + f\left(\frac{1-x}{x}\right)\right] \\
&= x^2\left[1 + \left(\frac{1-x}{x}\right)^2 f\left(\frac{x}{1-x}\right)\right] \\
&= x^2 + (1-x)^2 f\left(\frac{1-(1-x)}{1-x}\right) \\
&= x^2 + (1-x)^2 f\left(\frac{1}{1-x} - 1\right) \\
&= x^2 + (1-x)^2\left[f\left(\frac{1}{1-x}\right) - f(1)\right] \\
&= x^2 + (1-x)^2 f\left(\frac{1}{1-x}\right) - (1-x)^2 \\
&= x^2 + (1-x)^2\left[\left(\frac{1}{1-x}\right)^2 f(1-x)\right] - (1-x)^2
\end{aligned}$$

$$= x^2 + f(1-x) - (1-x)^2$$
$$= x^2 + f(1) - f(x) - (1-x)^2,$$

that is,

$$f(x) = x^2 + 1 - f(x) - (1-x)^2,$$

giving

$$2f(x) = 2x,$$

and

$$f(x) = x.$$

Hence the only such function is $f(x) = x$.

6. A sweater factory dyes its yarn in six different colors and each sweater is made with yarns of two different colors. If no two sweaters are made with the same pair of colors and each color is used in at least three sweaters, prove that some three of the sweaters contain all six colors between them.

Let G be a graph having a vertex for each color, a, b, c, d, e, f, with the edge (u, v) if and only if colors u and v are paired together in a sweater. Thus each edge represents a sweater and we would like to show that some three edges of G contain all six vertices between them.

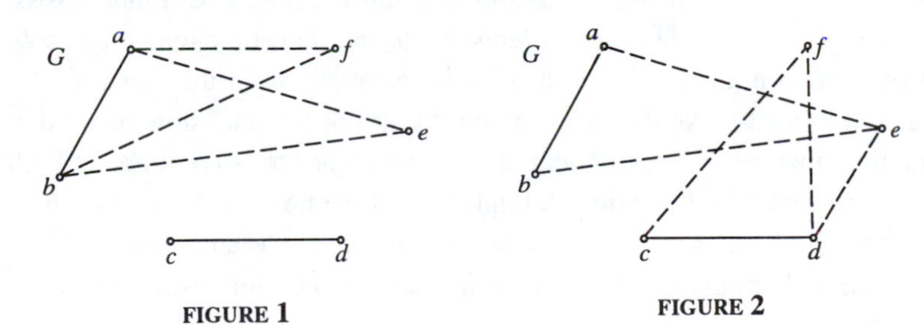

FIGURE 1 FIGURE 2

Since each color occurs in at least three sweaters, there are at least three edges at each vertex of G. Without loss of generality, suppose (a, b) is one of the edges (Figure 1). Now, two of the edges at c might go to a and b, but a third one must join c to some other vertex, say d. Consequently, if (e, f) is also an edge, then the three edges (a, b), (c, d), (e, f) contain all the vertices between them.

Suppose, then, that (e, f) is **not** an edge. This forces each of e and f to be joined to at least three of a, b, c, d. Thus, if e is not joined to both a and b, it would have to be joined to both c and d, and conversely; similarly for f. Hence the following two cases cover all the possibilities:

(i) e and f are both joined to the same pair, say both to a and b (Figure 1),

(ii) e and f are joined to different pairs, say e is joined to a and b and f is joined to c and d (Figure 2).

In case (i), the three edges (a, f), (b, e), and (c, d) clearly contain all the vertices.

In case (ii), we need to observe that, since (e, f) is not an edge, a third edge from e must go either to c or d. Since the alternatives are equivalent, suppose e is joined also to d (Figure 2). Then (a, b), (c, f), and (d, e) contain all the vertices and our argument is complete.

7. Is it possible to construct, from a set M having 22,222 elements, a collection of 50 subsets, $M_1, M_2, \ldots, M_{50}$, each containing 1111 elements, so that each two of the subsets have exactly 22 elements in common?

I encountered this problem some time back and wrote up a solution on a piece of foolscap which remained undisturbed in a folder of miscellaneous items until a few weeks ago. Looking it over at that time, I decided against including it in any of my collections and threw it away. Coming across the problem a couple of days ago in *Junge Mathematiker*, I reckoned it was worth reconsideration and sat down to solve it again. When in this position, one is tempted to try to take the easy way out by trying to recall how one did it the first time instead of working up the drive necessary to solve it again. Perhaps my heart wasn't in it, but do you think I could see how to go about a solution? Since the problem is really so simple, I include it here as an example of an easy problem with an inordinate potential for leading the mind astray in confusing and irrelevant considerations.

The solution is obtained immediately from a straightforward review of the subsets, noting in each case the **minimum number of new elements** the subset must bring into the fold.

M_1 starts things off with **1111** elements.

M_2 has 22 elements that occur also in M_1 and $(\mathbf{1111} - \mathbf{22})$ new elements.

> M_3 has 22 elements that occur in M_1 and 22 elements of M_2, and even if these 44 elements are all different, it must introduce $(\mathbf{1111 - 2 \cdot 22})$ new elements.
>
> M_4 has 22 elements in common with each of M_1, M_2, and M_3, but it can't avoid introducing a minimum of $(\mathbf{1111 - 3 \cdot 22})$ new elements.
>
> ....
>
> M_{50} must bring in a minimum of $(\mathbf{1111 - 49 \cdot 22})$ new elements.

Thus in order for the proposed division to be possible, the **minimum** number of elements in M must be

$$\begin{aligned}1111 &+ (1111 - 22) + (1111 - 2 \cdot 22) + \cdots + (1111 - 49 \cdot 22)\\ &= 50 \cdot 1111 - 22(1 + 2 + 3 + \cdots + 49) = 28{,}600.\end{aligned}$$

Thus a set M having a mere 22,222 elements is far too small for the proposal.

8. A group of 11 scientists wants to design a cabinet in which to keep their top secret papers. They propose to outfit the cabinet with a set of locks and supply a certain number of keys to each scientist so that the cabinet can be opened only when a majority of the scientists is present, that is,
   (i) no group of 5 or fewer is to be able to open all the locks, and
   (ii) every group of 6 or more is to have keys for all the locks.

   What is the minimum number of locks that must be built into the cabinet and what is the minimum number of keys each scientist must be given?

Clearly there must be a lock L_{ABCDE} which the 5 scientists $\{A, B, C, D, E\}$ cannot open, and also a lock L_{ABCDF} which keeps out the minority $\{A, B, C, D, F\}$. But these two locks can't be the same, for F has a key for L_{ABCDE} but not for L_{ABCDF}:

> the majority $\{A, B, C, D, E, F\}$ can open all the locks, but none of A, B, C, D, E is able to open L_{ABCDE}; thus it must be F who has the key to this lock.

Thus there needs to be a different lock L_{PQRST} for each subset $\{P, Q, R, S, T\}$ of 5 of the scientists. Therefore the project would require

$$\binom{11}{5} = 462 \text{ different locks,}$$

and since each scientist must carry a key for each subset of 5 other scientists (which his presence converts from a minority to a majority, thus requiring him to supply the key to the lock they can't open), each scientist would be obliged to carry around

$$\binom{10}{5} = 252 \text{ different keys.}$$

All of which goes to show that some things that sound plausible are utterly impractical.

9. S is a set of n points in the plane, $n \geq 4$. If each four points of S determine a nondegenerate convex quadrilateral, prove that all n points of S determine a convex n-gon.

If some three points of S were to be collinear, the quadrilateral they determine with any fourth point would be degenerate. Therefore no three points of S can be collinear.

Now suppose some point P of S were to lie inside the convex hull H of S (Figure 3). If H is partitioned into triangles by drawing all the diagonals from a vertex Q, then, because no three points of S are collinear, P would lie in the interior of one of the triangles QLM. But then the quadrilateral determined by Q, L, M, and P would not be convex. Hence no point of S can lie inside H, implying the n points of S determine the convex hull H itself.

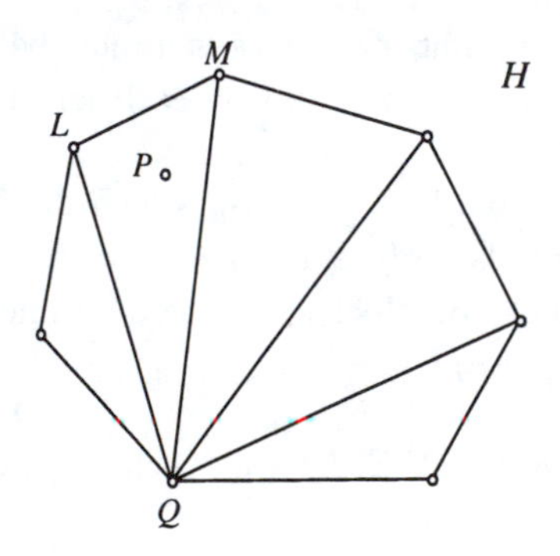

FIGURE 3

10. Finally, a problem with a glorious solution.

    S is any set of n points in the plane, $n \geq 2$. Let D and d, respectively, denote the greatest and least distances determined by two points of S. Prove that

$$D > \frac{\sqrt{3}}{2}(\sqrt{n} - 1)d.$$

Let C be the circle of minimum radius R that encloses the set S: one can imagine a large circle around S being shrunk until a tight contact is made with the outlying points of S. There are only two kinds of contact that can result:

(i) some two points P_i and P_j of S lie at the ends of a diameter of C, thus preventing further shrinking (Figure 4, Case (i)),

(ii) some three points of contact, P_i, P_j, P_k, determine an **acute** angled triangle (Figure 4, Case (ii)).

(An **obtuse** angled triangle, on its own, is not enough to put an end to the shrinking: the three vertices of an obtuse angled triangle must be contained in the interior of a semicircular arc (Figure 4, Case (iii)), implying that **more** than a semicircular arc does **not** make contact with S, in which case the circle can be pushed away from the points of contact and shrunk further.)

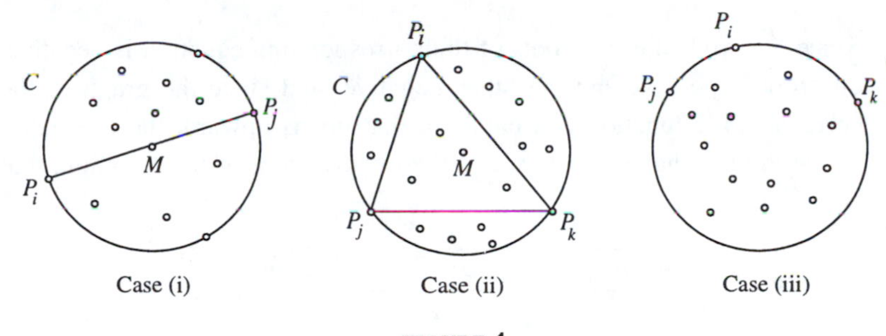

FIGURE 4

Now let a circle of radius $\frac{d}{2}$ be drawn about each point P_i of S as center (Figure 5).

These circles might touch each other, but no two of them can overlap in interior points without their centers being closer together than the minimum distance d. Of course, if a center is near or on the circumference of C, its little

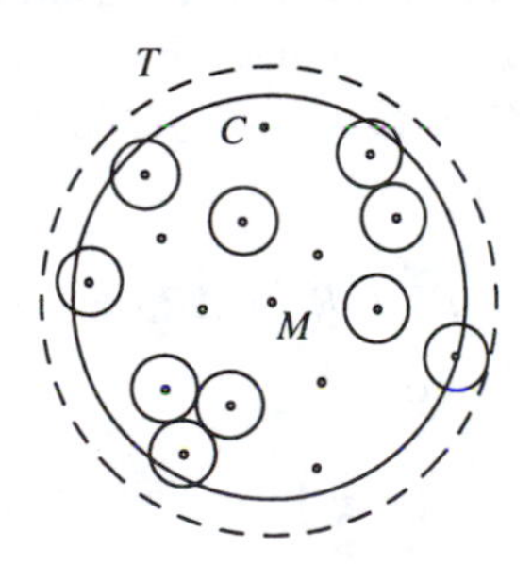

FIGURE 5

circle will protrude beyond C. However, a circle T of radius $R + \frac{d}{2}$, concentric with C, will contain all the little circles. In fact, because these non-overlapping little circles cannot be fitted together so as to cover T completely (there are always some little spaces between the circles), the area of T is strictly greater than the total area of the n little circles, and we have

$$\pi\left(R + \frac{d}{2}\right)^2 > n\pi\left(\frac{d}{2}\right)^2.$$

This leads to the result $R > (\sqrt{n} - 1)\frac{d}{2}$ as follows: We have

$$R^2 + dR + \frac{d^2}{4} > n \cdot \frac{d^2}{4}$$

$$R^2 + dR - (n-1)\frac{d^2}{4} > 0.$$

Since the product of the roots of the corresponding equation is negative, one root is positive and the other negative, and since the graph of the corresponding function is a parabola that opens upward, the inequality above implies that R must exceed the positive root of the equation, that is,

$$R > \frac{-d + \sqrt{d^2 + (n-1)d^2}}{2} = \frac{-d + d\sqrt{n}}{2} = (\sqrt{n} - 1)\frac{d}{2}.$$

Therefore, in case (i), we have

$$D = P_i P_j = 2R > 2\left[(\sqrt{n} - 1)\frac{d}{2}\right] > \sqrt{3}\left[(\sqrt{n} - 1)\frac{d}{2}\right],$$

as desired.

Since an acute angled triangle inscribed in a circle must contain the center of the circle in its interior (Figure 6), in case (ii), one of the sides $P_i P_k$ must subtend at the center M of C an angle $\theta \geq \frac{1}{3}(360°) = 120°$. Thus, by the law of cosines, we have

$$\begin{aligned} D^2 \geq P_i P_k^2 &= R^2 + R^2 - 2R \cdot R \cdot \cos\theta \\ &= 2R^2(1 - \cos\theta) \\ &\geq 2R^2\left[1 - \left(-\frac{1}{2}\right)\right] \quad \text{since } \cos\theta \leq -\frac{1}{2}, \\ &= 3R^2. \end{aligned}$$

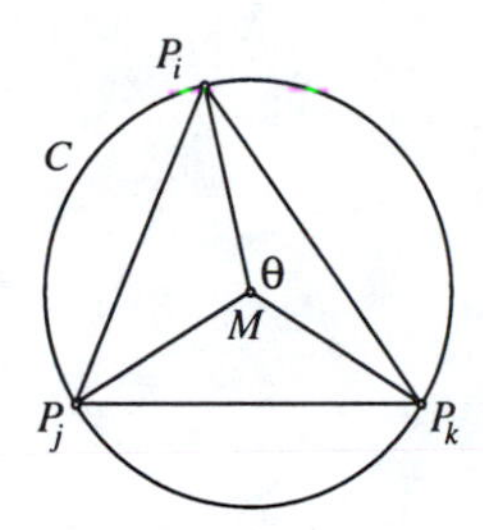

FIGURE 6

Hence

$$\begin{aligned} D &\geq \sqrt{3}R \\ &> \sqrt{3}\left[\left(\sqrt{n}-1\right)\frac{d}{2}\right] \quad \left(\text{recall } R > \left(\sqrt{n}-1\right)\frac{d}{2}\right) \\ &= \frac{\sqrt{3}}{2}\left(\sqrt{n}-1\right)d. \end{aligned}$$

SECTION 6

28 Problems from *Pi Mu Epsilon Journal*

(Specific references are given at the end.)

1. The arms of an equal-arm balance are never exactly the same length. To eliminate the error due to this unavoidable contingency, it is proposed that objects be weighed twice, once on each side of the balance, and the average taken. Show, however, that the average of the two weighings never gives the correct weight unless the arms are exactly the same length, and determine how the weighings ought to be treated in order to give the correct weight.

Let the fulcrum of the balance be F and the lengths of the arms AF and BF be a and b respectively; also, suppose an object of weight W weighs in at S and T when placed in pans A and B respectively (Figure 1).

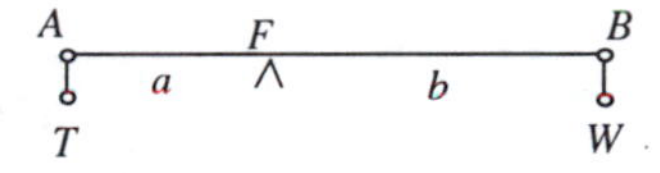

FIGURE 1

Then

$$aW = bS \quad \text{and} \quad aT = bW,$$

and dividing, we get

$$\frac{W}{T} = \frac{S}{W},$$

giving

$$W = \sqrt{ST},$$

the **geometric mean** of the weighings.

Since the geometric mean $\leq$ the arithmetic mean, with equality only when $S = T$, the average is always overweight unless $W = S$ and $a = b$.

2. Given a circle P and two points A and B outside P which are not in line with its centre O, construct a circle Q to go through A and B and divide the circumference of P into two equal arcs (Figure 2).

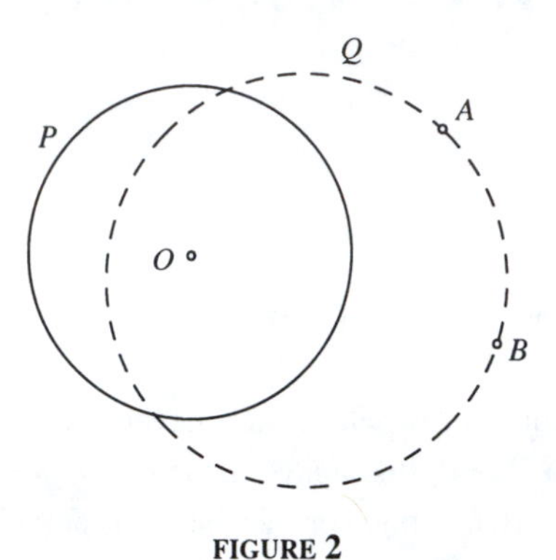

FIGURE 2

Let the radius of P be r. Suppose Q intersects P at E and F; then EF is a diameter of P. Let AO extended meet Q at C (Figure 3). Then AC and EF are intersecting chords in Q and we have

$$AO \cdot OC = EO \cdot OF = r^2,$$

giving

$$OC = \frac{r^2}{AO},$$

which can be determined from the given information. Therefore Q is obtained simply by marking C on AO extended and drawing the circle around $\triangle ABC$.

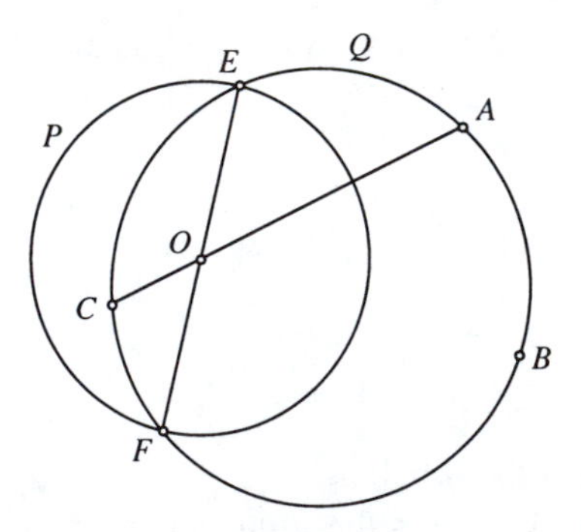

FIGURE 3

3. Solve the following equation for x in terms of the given positive real number c:

$$2\log_x c - \log_{cx} c - 3\log_{c^2x} c = 0.$$

If $\log_a x = y$, then $a^y = x$. Hence $a = x^{1/y}$ and $\log_x a = \frac{1}{y}$. That is to say, $\log_a x$ and $\log_x a$ are reciprocals. Thus the given equation is

$$\frac{2}{\log_c x} - \frac{1}{\log_c cx} - \frac{3}{\log_c c^2x} = 0.$$

Letting $\log_c x = k$, we have that $\log_c cx = \log_c x + \log_c c = k + 1$ and similarly, $\log_c c^2x = k + 2$, and the equation becomes

$$\frac{2}{k} - \frac{1}{k+1} - \frac{3}{k+2} = 0,$$
$$2(k^2 + 3k + 2) - (k^2 + 2k) - 3(k^2 + k) = 0,$$
$$2k^2 - k - 4 = 0,$$

and

$$k = \frac{1 \pm \sqrt{33}}{4}.$$

Hence

$$x = c^k = c^{(1\pm\sqrt{33})/4}.$$

4. Let T_n be the right triangle with sides of lengths $(4n^2, 4n^4 - 1, 4n^4 + 1)$, n a positive integer, and let α_n be the angle opposite the side of length $4n^2$ (Figure 4).

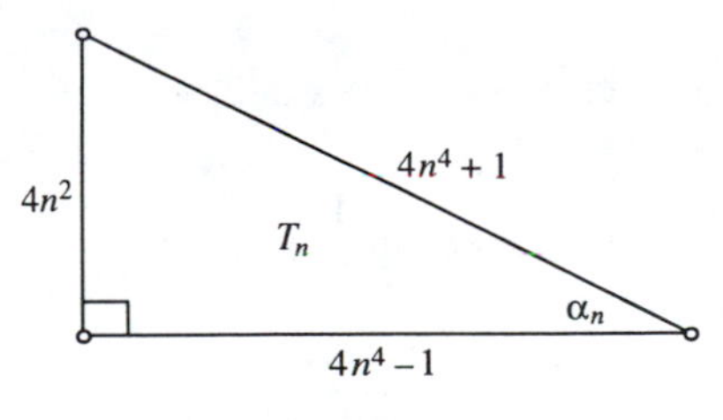

FIGURE 4

Prove that, as n runs through the positive integers, the sum S of all the angles α_n is a right angle:

$$S = \alpha_1 + \alpha_2 + \cdots = 90°.$$

Clearly

$$\tan \alpha_n = \frac{4n^2}{4n^4 - 1}, \quad \text{and so} \quad \alpha_n = \arctan \frac{4n^2}{4n^4 - 1},$$

and the problem can be seen as summing an infinite series of arctangents.

Let us begin by recalling that

$$\arctan x + \arctan y = \arctan \frac{x+y}{1-xy}$$

(the tangent of each side is $\frac{x+y}{1-xy}$).

In particular, when $x = y$ we have

$$\arctan \frac{2x}{1-x^2} = 2 \cdot \arctan x.$$

Now, dividing the numerator and denominator of $\frac{4n^2}{4n^4-1}$ by $4n^4$, we get

$$\frac{4n^2}{4n^4 - 1} = \frac{\frac{1}{n^2}}{1 - \frac{1}{4n^4}} = \frac{2\left(\frac{1}{2n^2}\right)}{1 - \left(\frac{1}{2n^2}\right)^2}.$$

Therefore

$$\begin{aligned} \alpha_n &= \arctan \frac{2\left(\frac{1}{2n^2}\right)}{1 - \left(\frac{1}{2n^2}\right)^2} \\ &= 2 \arctan \frac{1}{2n^2}, \end{aligned}$$

and

$$\begin{aligned} S &= 2 \sum_{i=1}^{\infty} \arctan \frac{1}{2i^2} \\ &= 2\left(\arctan \frac{1}{2} + \arctan \frac{1}{8} + \arctan \frac{1}{18} + \cdots\right). \end{aligned}$$

From

$$\arctan x + \arctan y = \arctan \frac{x+y}{1-xy},$$

we have

$$\arctan \frac{1}{2} + \arctan \frac{1}{8} = \arctan \frac{5/8}{15/16} = \arctan \frac{2}{3};$$

that is, for $n = 2$,

$$\sum_{i=1}^{n} \left(\arctan \frac{1}{2i^2} \right) = \arctan \frac{n}{n+1}.$$

It is easy to show by induction that this is generally true:

The result clearly holds for $n = 1$. Now, suppose that

$$S_{n-1} = \sum_{i=1}^{n-1} \left(\arctan \frac{1}{2i^2} \right) = \arctan \frac{n-1}{n}.$$

Then

$$\begin{aligned}
S_n = \sum_{i=1}^{n} \left(\arctan \frac{1}{2i^2} \right) &= S_{n-1} + \arctan \frac{1}{2n^2} \\
&= \arctan \frac{n-1}{n} + \arctan \frac{1}{2n^2} \\
&= \arctan \frac{\frac{n-1}{n} + \frac{1}{2n^2}}{1 - \left(\frac{n-1}{n}\right)\left(\frac{1}{2n^2}\right)} \\
&= \arctan \frac{2n^2(n-1) + n}{2n^3 - (n-1)}.
\end{aligned}$$

It remains to show that

$$\frac{2n^2(n-1) + n}{2n^3 - (n-1)} = \frac{n}{n+1},$$

that is, that

$$2n^2(n^2 - 1) + n^2 + n = 2n^4 - n^2 + n,$$

which is clearly so.

Therefore the desired sum is given by

$$\begin{aligned}
S &= \lim_{n \to \infty} \left(2 \arctan \frac{n}{n+1} \right) \\
&= 2 \lim_{n \to \infty} \arctan \frac{n}{n+1} \\
&= 2(\arctan 1) \\
&= 2(45^\circ) \\
&= 90^\circ.
\end{aligned}$$

5. M is an arbitrary point on the median AA' of $\triangle ABC$ and CM meets AB at N (Figure 5). Prove that the ratio in which AA' is divided by M is twice the ratio in which AB is divided by N:

$$\frac{AM}{MA'} = 2 \cdot \frac{AN}{NB}.$$

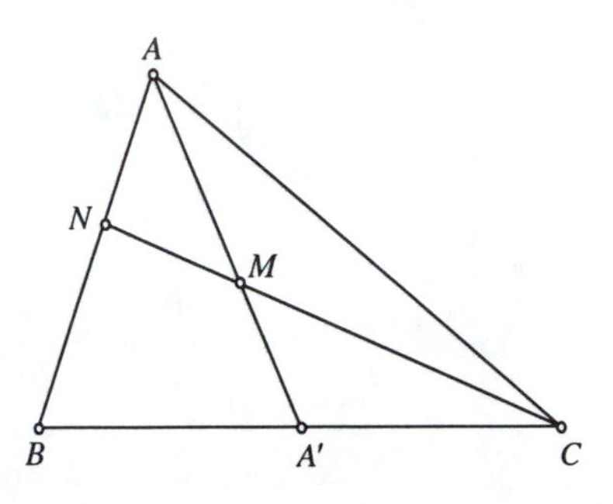

FIGURE 5

Let the straight line through A' parallel to CMN meet AB at D (Figure 6). Then, in $\triangle BCN$, A' bisects BC and $A'D$ is parallel to CN, implying that D is the midpoint of BN.

Now, in $\triangle ADA'$, NM is parallel to DA', and so

$$\frac{AM}{MA'} = \frac{AN}{ND},$$

and since $ND = \frac{1}{2}NB$, we have

$$\frac{AM}{MA'} = 2 \cdot \frac{AN}{NB}.$$

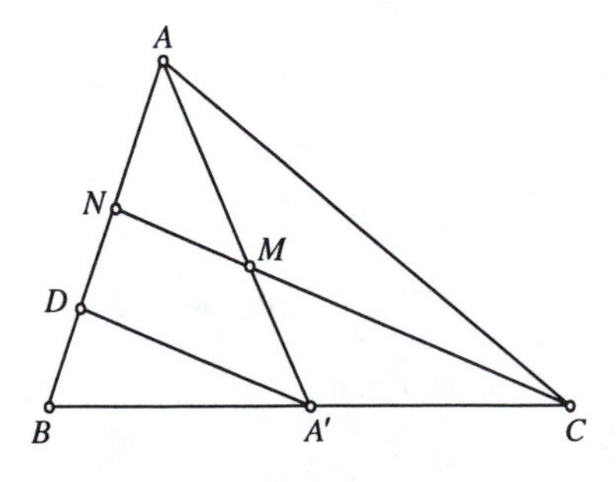

FIGURE 6

6. Let p_n denote the nth prime number. It is customary to denote the **number** of primes $\leq m$ by $\pi(m)$; e.g. $\pi(12) = 5$ (due to 2, 3, 5, 7, 11).

For all $n \geq 6$, prove that

$$\pi(\sqrt{p_1 p_2 \cdots p_n}) > 2n.$$

Proceeding by induction, for $n = 6$, we have

$$\pi(\sqrt{p_1 p_2 p_3 p_4 p_5 p_6}) = \pi(\sqrt{30030}) \geq \pi(50) = 15 > 2 \cdot 6,$$

validating the claim for $n = 6$.

Now suppose

$$\pi(\sqrt{p_1 p_2 \cdots p_n}) > 2n \quad \text{for some } n \geq 6.$$

We wish to show that

$$\pi(\sqrt{p_1 p_2 \cdots p_n p_{n+1}}) > 2n + 2,$$
$$\text{i.e., } \pi(\sqrt{p_{n+1}}\sqrt{p_1 p_2 \cdots p_n}) > 2n + 2.$$

Since $n \geq 6$, then $p_{n+1} \geq p_7 = 17$, and $\sqrt{p_{n+1}} > 4$, and

$$\pi(\sqrt{p_{n+1}}\sqrt{p_1 p_2 \cdots p_n}) \geq \pi(4\sqrt{p_1 p_2 \cdots p_n}).$$

Now, Bertrand's Postulate guarantees a prime number between k and $2k$ for all $k > 1$. Thus

$$[\pi(4k) - \pi(2k)] \text{ and } [\pi(2k) - \pi(k)] \text{ are each at least } 1,$$

and

$$[\pi(4k) - \pi(2k)] + [\pi(2k) - \pi(k)] \geq 1 + 1.$$

Observing that $\pi(4k) = [\pi(4k) - \pi(2k)] + [\pi(2k) - \pi(k)] + \pi(k)$, then

$$\begin{aligned}
\pi(\sqrt{p_1 p_2 \cdots p_n p_{n+1}}) &\geq \pi(4\sqrt{p_1 p_2 \cdots p_n}) \\
&= [\pi(4\sqrt{p_1 p_2 \cdots p_n}) - \pi(2\sqrt{p_1 p_2 \cdots p_n})] \\
&\quad + [\pi(2\sqrt{p_1 p_2 \cdots p_n}) - \pi(\sqrt{p_1 p_2 \cdots p_n})] \\
&\quad + \pi(\sqrt{p_1 p_2 \cdots p_n}) \\
&> 1 + 1 + 2n \qquad (\text{since } \pi(\sqrt{p_1 p_2 \cdots p_n}) > 2n) \\
&= 2n + 2, \qquad \text{as required.}
\end{aligned}$$

7. (a) In $\triangle ABC$, $\angle A = 30°$ and $\angle B = 45°$. Find k such that $c^2 = ka^2 + b^2$.
   (b) In an arbitrary triangle ABC, find k such that $c^2 = ka^2 + b^2$.

(a) (Solution by Ian McGee and Lloyd Auckland, University of Waterloo.)

Let the altitude CD be taken as the unit of length (Figure 7). Then $\triangle ACD$ is a 30°-60°-90° triangle with sides $AC = 2$ and $AD = \sqrt{3}$; also, $\triangle BCD$ is right angled and isosceles with sides $CD = BD = 1$ and $BC = \sqrt{2}$. Therefore

$$c^2 = ka^2 + b^2,$$

requires

$$(\sqrt{3}+1)^2 = k \cdot 2 + 4,$$
$$4 + 2\sqrt{3} = 2k + 4,$$

giving

$$k = \sqrt{3}.$$

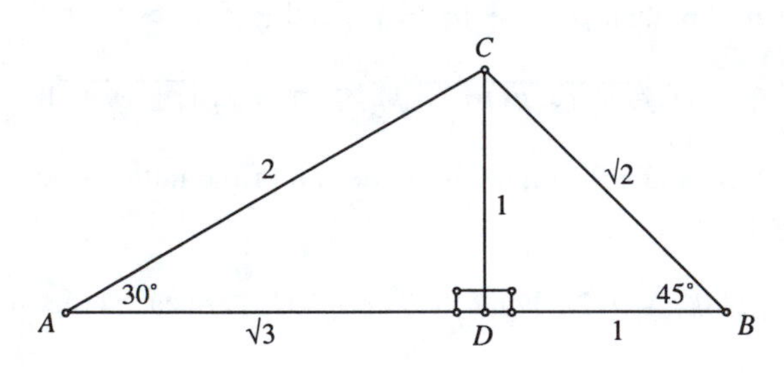

FIGURE 7

(b) By the law of cosines

$$c^2 = b^2 + a^2 - 2ab \cdot \cos C$$
$$= b^2 + a^2 \left[1 - (2\cos C)\frac{b}{a}\right],$$

implying

$$k = 1 - (2\cos C)\frac{b}{a}.$$

Since $\dfrac{b}{a} = \dfrac{\sin B}{\sin A}$ by the law of sines, then

$$\begin{aligned} k &= 1 - (2\cos C)\frac{\sin B}{\sin A} \\ &= \frac{\sin A - 2\sin B\cos C}{\sin A}, \end{aligned}$$

and since $\sin A = \sin(180^\circ - A) = \sin(B + C)$, we have

$$\begin{aligned} k &= \frac{\sin(B + C) - 2\sin B\cos C}{\sin A} \\ &= \frac{\sin B\cos C + \cos B\sin C - 2\sin B\cos C}{\sin A} \\ &= \frac{\sin C\cos B - \cos C\sin B}{\sin A} \\ &= \frac{\sin(C - B)}{\sin A}. \end{aligned}$$

8. AOB is a quarter of a circle with center O (Figure 8). Prove that the incircle of $\triangle AOB$ has twice the radius of the largest circle that can be inscribed in the minor segment determined by chord AB.

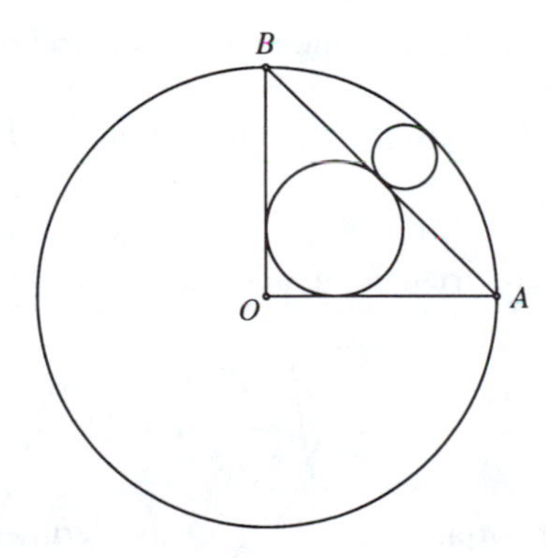

FIGURE 8

Clearly the centers of the two circles in question lie on the bisector OC of $\angle AOB$, which is also the altitude from O in $\triangle AOB$ (Figure 9). If radius OA is taken as the unit of length, then $OC = 1$ and the the length of the altitude $OD = \sin 45^\circ = \frac{1}{\sqrt{2}}$. Thus

$$\text{diameter } CD = 1 - \frac{1}{\sqrt{2}} = \frac{1}{2}(2 - \sqrt{2}).$$

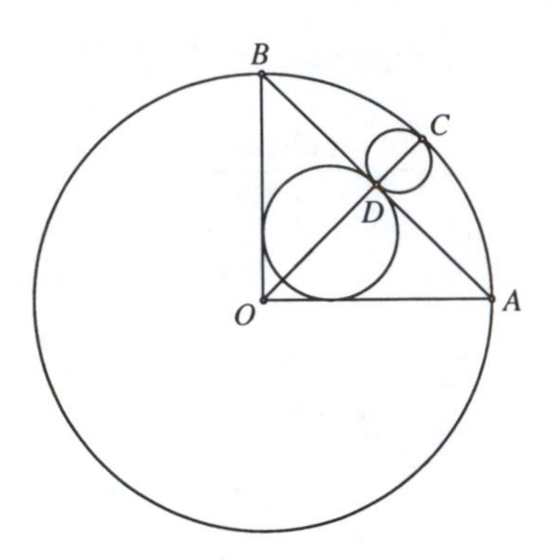

FIGURE 9

It remains only to recall that the diameter of the incircle of a right triangle is given by (the sum of the legs – the hypotenuse); for $\triangle AOB$, then,

$$\text{the in-diameter} = 1 + 1 - \sqrt{2} = 2 - \sqrt{2} = 2CD.$$

9. Prove that the perimeter of every triangle of area $\frac{1}{\pi}$ exceeds 2.

Solution 1 Unfortunately this problem loses much of its glamour since it follows immediately from the well known isoperimetric result

> *of all triangles with a given area, the one having smallest perimeter is equilateral.*

The area of the equilateral triangle of side s is $\frac{s^2\sqrt{3}}{4}$, and with $\frac{s^2\sqrt{3}}{4} = \frac{1}{\pi}$, we have

$$s = \sqrt{\frac{4}{\pi\sqrt{3}}},$$

giving the perimeter of a triangle of area $\frac{1}{\pi} \geq$ the perimeter of the equilateral triangle of area $\frac{1}{\pi}$

$$= 3s = \sqrt{\frac{12\sqrt{3}}{\pi}} > \sqrt{5} > 2.$$

Solution 2 From the triangle formula $\Delta = rs$, where Δ is the area, r the inradius, and s the semiperimeter (i.e., $s = \frac{p}{2}$, where p is the perimeter), we have

$$\frac{1}{\pi} = r\frac{p}{2},$$

and

$$p = \frac{2}{r\pi}.$$

However, since the area of the incircle of a triangle is obviously less than the area of the triangle itself, we have

$$\pi r^2 < \frac{1}{\pi} \quad \text{and} \quad \pi^2 r^2 < 1, \text{ yielding } \pi r < 1$$

and making

$$p = \frac{2}{r\pi} > 2.$$

Solution 3 It has been established in many places that the triangle ABC of minimum perimeter which has a given base $BC = a$ and a given area (i.e., a given altitude h to base BC) is the **isosceles** triangle $A'BC$ having $A'B = A'C$ (Figure 10, where $AA' \| BC$ and B' is the reflection of B in AA').

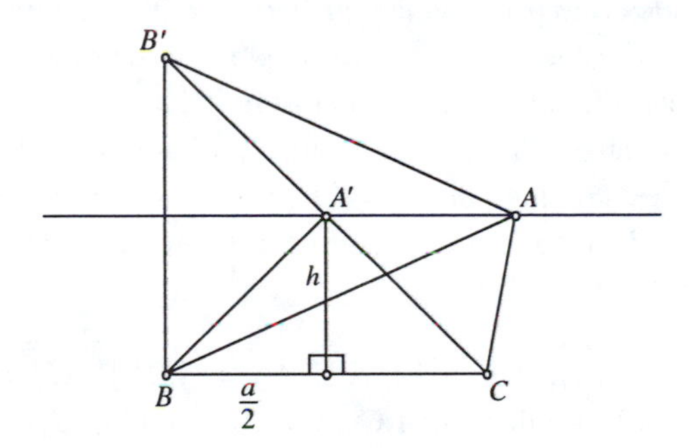

FIGURE 10

Thus every triangle ABC of area $\frac{1}{\pi}$ that has base $BC = a$ has a perimeter which is at least as great as the perimeter of the corresponding isosceles triangle $A'BC$. Since $\frac{1}{2}ah = \frac{1}{\pi}$, then $h = \frac{2}{\pi a}$, and we have

$$\begin{aligned}
&\text{the perimeter of } \triangle ABC \\
&\quad \geq \text{the perimeter of } \triangle A'BC \\
&\quad = a + 2\sqrt{\frac{a^2}{4} + h^2} = a + 2\sqrt{\frac{a^2}{4} + \frac{4}{\pi^2 a^2}} = a + \sqrt{a^2 + \frac{16}{\pi^2 a^2}} \\
&\quad = a + \frac{1}{a}\sqrt{a^4 + \frac{16}{\pi^2}} > a + \frac{1}{a}\sqrt{a^4 + 1} > a + \frac{1}{a} \geq 2 \quad \text{for all } a > 0.
\end{aligned}$$

10. Recall that Euler's function $\varphi(n)$ counts the number of positive integers m which are less than or equal to the positive integer n and are relatively prime to n; for example, $\varphi(10) = 4$, for the numbers 1, 3, 7, 9.

 Prove that the φ-function enjoys the remarkable property

$$\frac{\varphi(1)}{1} + \frac{\varphi(1+1)}{11} + \frac{\varphi(1+1+1)}{111} + \cdots = 1.111111\ldots;$$

 in fact, show that this is true no matter in what scale the integers 1, 11, 111, ... and the "decimal" 1.1111 ... might be expressed.

Recall that (i) $(1-x)^{-1} = 1 + x + x^2 + \cdots$ and (ii) $(1-x)^{-2} = 1 + 2x + 3x^2 + \cdots$. Besides these expansions, the solution uses the remarkable property of the φ-function

$$\text{(iii)} \qquad \sum_{d|n} \varphi(d) = n.$$

In order to make our solution essentially self-sufficient, let us digress briefly to consider Gauss's brilliant proof of this result.

The basic idea is to partition the integers $1, 2, 3, \ldots, n$ into classes having sizes given by the values $\varphi(d)$, where d is a divisor of n; clearly, the total number of members in all the classes is just n, itself.

Suppose the integers $\{1, 2, \ldots, n\}$ are distributed into classes C_d according to their greatest common divisors d with n, that is, m is put into C_d if and only if $(m, n) = d$. If d is not a divisor of n, C_d remains empty. For $n = 12$, for example, we have

$C_1 = \{1, 5, 7, 11\}$, $C_2 = \{2, 10\}$, $C_3 = \{3, 9\}$, $C_4 = \{4, 8\}$, $C_6 = \{6\}$, $C_{12} = \{12\}$, (with each of C_5, C_7, C_8, C_9, C_{10}, C_{11} being empty).

Now we ask "How many integers m are there in C_d?". In order to have $(m, n) = d$, we must have, for some **relatively prime** positive integers k and t, that

$$m = kd, \quad \text{and} \quad n = td;$$

and since $m \le n$, then $k \le t$. That is to say, for each member m of C_d, there is a positive integer k which is $\le t$ and relatively prime to t.

Conversely, each such k determines a member $m = kd$ of C_d:

$$k \le t \quad \text{implies} \quad m = kd \le td = n,$$

and

$$(k, t) = 1 \quad \text{gives} \quad (m, n) = (kd, td) = d(k, t) = d \cdot 1 = d.$$

Thus the cardinality of C_d is $\varphi(t)$, that is, $\varphi\left(\frac{n}{d}\right)$, and altogether we have

$$\sum_{d|n} \varphi\left(\frac{n}{d}\right) = n.$$

Finally, we need only observe that $\frac{n}{d}$ is the complementary divisor of d, and that if the divisors of n are $d_1 < d_2 < \cdots < d_{r-1} < d_r$, then

$$\frac{n}{d_1} > \frac{n}{d_2} > \cdots > \frac{n}{d_{r-1}} > \frac{n}{d_r}$$

is the same set of numbers in reverse order. Hence

$$\sum_{d|n} \varphi(d) = \sum_{d|n} \varphi\left(\frac{n}{d}\right) = n.$$

Now we can proceed comfortably with the solution.

Let the base of the number scale be b. Then the number $111\ldots1$, containing r 1's, is given by

$$b^{r-1} + b^{r-2} + \cdots + b + 1 = \frac{b^r - 1}{b - 1}.$$

Thus

$$\sum_{r=1}^{\infty} \frac{\varphi(1 + 1 + \cdots + 1)}{11\ldots1}, \quad \text{with } r \text{ 1's top and bottom,}$$

$$\begin{aligned}
&= \sum_{r=1}^{\infty} \frac{(b-1)\cdot\varphi(r)}{b^r - 1} \\
&= (b-1)\sum_{r=1}^{\infty} \frac{\varphi(r)}{b^r\cdot(1 - b^{-r})} \\
&= (b-1)\sum_{r=1}^{\infty} \varphi(r)\cdot b^{-r}(1 - b^{-r})^{-1} \\
&= (b-1)\sum_{r=1}^{\infty} \varphi(r)\cdot b^{-r}(1 + b^{-r} + b^{-2r} + \cdots) \\
&= (b-1)\sum_{r=1}^{\infty} \varphi(r)\cdot(b^{-r} + b^{-2r} + b^{-3r} + \cdots) \\
&= (b-1)[\varphi(1)\cdot(b^{-1} + b^{-2} + b^{-3} + \cdots) \\
&\qquad + \varphi(2)\cdot(b^{-2} + b^{-4} + b^{-6} + \cdots) \\
&\qquad + \varphi(3)\cdot(b^{-3} + b^{-6} + b^{-9} + \cdots)
\end{aligned}$$

$$
\begin{aligned}
&+\varphi(4)\cdot(b^{-4}+b^{-8}+b^{-12}+\cdots)\\
&+\varphi(5)\cdot(b^{-5}+b^{-10}+b^{-15}+\cdots)\\
&+\varphi(6)\cdot(b^{-6}+b^{-12}+b^{-18}+\cdots)\\
&+\cdots].
\end{aligned}
$$

Neglecting the initial factor $(b-1)$ for the moment, when the large bracket is multiplied out, each term is of the form $\varphi(k)\cdot b^{-mk}$ and is therefore a term in b^{-n} if and only if $n = mk$, that is, if and only if k is a **divisor** of n; for example, the total term in b^{-6} is the sum of the terms $\varphi(k)\cdot b^{-6}$ where k is a divisor of 6,

$$\varphi(1)\cdot b^{-6}+\varphi(2)\cdot b^{-6}+\varphi(3)\cdot b^{-6}+\varphi(6)\cdot b^{-6},$$

and in general the coefficient of the term in b^{-n} is

$$\sum_{d|n}\varphi(d), \qquad \text{which is simply } n, \text{ itself.}$$

Hence the series is given by

$$
\begin{aligned}
(b-1)\sum_{n=1}^{\infty} nb^{-n} &= (b-1)[b^{-1}+2(b^{-1})^2+3(b^{-1})^3+\cdots]\\
&= (b-1)\cdot b^{-1}[1+2(b^{-1})+3(b^{-1})^2+\cdots]\\
&= (1-b^{-1})\cdot(1-b^{-1})-2\\
&= (1-b^{-1})^{-1}\\
&= 1+b^{-1}+b^{-2}+b^{-3}+\cdots\\
&= 1+\frac{1}{b}+\frac{1}{b^2}+\cdots\\
&= 1.1111\ldots \text{ in base } b.
\end{aligned}
$$

11. If finite real numbers x, y, z, t satisfy the system of equations

$$
\begin{aligned}
x(yz+zt+ty)&=0,\\
y(xz+xt-zt)&=0,\\
z(xt+xy-ty)&=0,\\
t(xy+xz-yz)&=0,
\end{aligned}
$$

prove that at least one of x, y, z, t must be equal to zero; in fact, show that at least two of them must be zero.

If none of the variables is zero, the equations can be satisfied only if each of the bracketed factors is zero. In this case, if each bracket is divided by the product of the three variables it contains, we would have

$$\frac{1}{t}+\frac{1}{y}+\frac{1}{z}=0, \tag{1}$$

$$\frac{1}{t}+\frac{1}{z}-\frac{1}{x}=0, \tag{2}$$

$$\frac{1}{y}+\frac{1}{t}-\frac{1}{x}=0, \tag{3}$$

$$\frac{1}{z}+\frac{1}{y}-\frac{1}{x}=0. \tag{4}$$

Adding (2), (3), and (4), and using (1), we get $-\frac{3}{x}=0$, which is impossible. Hence at least one of the variables must be zero.

Now, if $x=0$, then the equation $y(xz+xt-zt)=0$ yields $-yzt=0$, implying a second variable must also be zero; similarly for the other cases.

12. Recall that the arithmetic function $\tau(n)$ counts the number of positive divisors of the positive integer n and that the Möbius function $\mu(n)$ is defined as follows:

    Let the prime decomposition of n be $p_1^{a_1}p_2^{a_2}\cdots p_k^{a_k}$;

then	$\mu(1)=1$,	
and for $n\geq 2$,	$\mu(n)=0$	if any exponent $a_i>1$;
otherwise	$\mu(n)=(-1)^k$	(i.e., when $n=p_1p_2\cdots p_k$).

For example, for a prime p, $\mu(p)=-1$, $\mu(12)=\mu(2^2\cdot 3)=0$, and $\mu(30)=\mu(2\cdot 3\cdot 5)=(-1)^3=-1$.

Prove that

$$\tau(n)+\mu^2(n)=\tau(n^2)$$

if and only if n is a prime number.

(a) If n is a prime p, then

$$\tau(n)+\mu^2(n)=\tau(p)+\mu^2(p)=2+(-1)^2=3=\tau(p^2),$$

satisfying the given condition.

(b) Suppose $\tau(n) + \mu^2(n) = \tau(n^2)$. Since the condition is not satisifed by $n = 1$ (we have $\tau(1) + \mu^2(1) = 1 + 1$, while $\tau(1^2) = 1$), then n cannot be 1.

Now, suppose the prime decomposition of n is

$$n = p_1^{a_1} p_2^{a_2} \cdots p_k^{a_k}.$$

If any exponent $a_i > 1$, then $\mu(n) = 0$, in which case the condition reduces to

$$\tau(n) = \tau(n^2).$$

Since n^2 has more divisors than n except for $n = 1$, this implies $n = 1$. But we saw above that n cannot be 1. It follows then, that all the exponents $a_i = 1$, and

$$n = p_1 p_2 \cdots p_k, \qquad n^2 = p_1^2 p_2^2 \cdots p_k^2,$$

in which case the given condition yields

$$\tau(n) + \mu^2(n) = 2^k + (-1)^{2k} = \tau(n^2) = 3^k, \quad \text{i.e., } 2^k + 1 = 3^k.$$

Clearly this holds only for $k = 1$, making $n = p_1$, a prime.

13. A pentagon $ABCDE$ is inscribed in a circle to have sides AB, BC, and DE of length equal to the radius r; the sides CD and EA may be any lengths so long as the pentagon doesn't cross itself (Figure 11).

    Prove that the midpoints F and G of CD and AE always determine an equilateral triangle with vertex B.

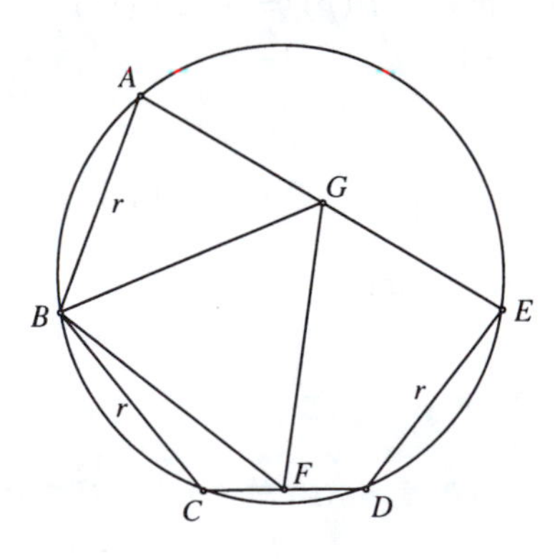

FIGURE 11

Let O be the center of the circle (in order to avoid clutter, O is not labelled in Figure 12). Since the segment from the center of a circle to the midpoint of a chord is perpendicular to the chord, then OF is perpendicular to CD, and since $OC = OD$, then OF is the bisector of $\angle COD$ in isosceles triangle COD. Let $\angle COF = \angle DOF = x$, and similarly, let $\angle AOG = \angle EOG = y$.

With the three 60° angles in the equilateral triangles AOB, BOC, and DOE, the sum of all the angles at O is

$$3(60^\circ) + 2x + 2y = 360^\circ$$

giving $x + y = 90^\circ$.

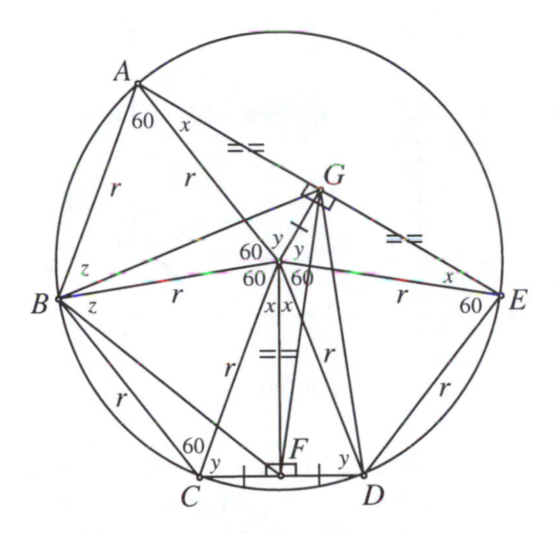

FIGURE 12

In right triangles COF and DOF, then, the angles at C and D are y; similarly, angle x occurs in $\triangle AOE$ at A and E. Then $\triangle AOG \equiv \triangle OCF$ (AAS), implying $AG = OF$ and $OG = CF$.

Now let us prove that $BG = BF$ by showing each is equal to GD:

Triangles ABG and GED are congruent (SAS), implying $BG = GD$. Similarly, $\triangle BCF \equiv \triangle DOG$, giving $BF = GD$.

Thus $\triangle BFG$ is at least isosceles, and it remains to show that the angle between the equal sides is 60°.

Now,

$$\triangle ABG \equiv \triangle BOF \text{ (SSS)},$$

and so $\angle ABG = \angle OBF$ ($= z$ in the figure).

Thus, adding $\angle OBG$ to each of these angles, we get the desired

$$\angle FBG = \angle ABO = 60^\circ.$$

14. A unit square $ABCD$ and its inscribed circle are given (Figure 13). L and M are the midpoints of the sides CD and AD. A point P is chosen on CL, and AQ is drawn parallel to MP to meet side BC at Q. Prove, for all choices of P, that PQ is tangent to the incircle.

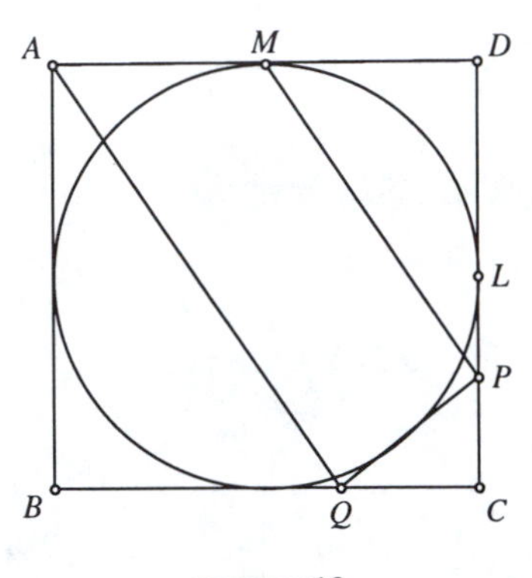

FIGURE 13

Let N be the midpoint of BC and let the lengths of BQ and DP be x and y (Figure 14). Clearly the angles α at Q, A, and M are equal due to pairs of parallel lines. Thus the tangents of these angles at Q and M are equal and we

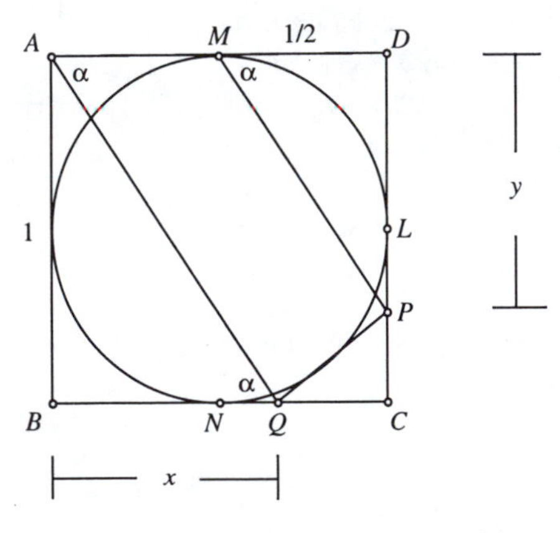

FIGURE 14

have

$$\frac{1}{x} = \frac{y}{1/2}, \quad \text{giving} \quad 2xy = 1.$$

Now

$$\begin{aligned} PQ^2 &= QC^2 + PC^2 \\ &= (1-x)^2 + (1-y)^2 \\ &= x^2 + y^2 + 1 + 1 - 2x - 2y, \end{aligned}$$

and since $2xy = 1$, then

$$\begin{aligned} PQ^2 &= x^2 + y^2 + 1 + 2xy - 2x - 2y \\ &= (x + y - 1)^2, \end{aligned}$$

implying

$$\begin{aligned} PQ &= x + y - 1 \\ &= \left(x - \frac{1}{2}\right) + \left(y - \frac{1}{2}\right) \\ &= QN + PL. \end{aligned}$$

But QN and PL are tangents to the incircle, and therefore the length of PQ is equal to the sum of the lengths of the tangents from P and Q. It is easy to deduce a contradiction, then, unless PQ is also tangent to the circle.

It is tempting to stop our solution at this point on the basis that this final step needs no elaboration. However, it isn't obvious and I am always happier when all the loose ends are tied up. At least its justification requires only the following simple arguments.

Let S and T be the points of contact of the tangents from P and Q, respectively, that lie in the interior of the square. We would like to show that S and T are the same point, for then each of PS and QT, being perpendicular to the radius to $S\ (= T)$, would be an extension of the other, implying PQ is also tangent to the circle (Figure 15).

Now, if S and T are not the same point, then the segments PS and QT either (i) cross at a point R which is interior to both of them (Figure 16a), or (ii) they don't intersect at all (Figure 16b). In case (i), we have

$$PS + QT > PR + QR > PQ, \quad \text{contradicting} \quad PS + QT = PQ,$$

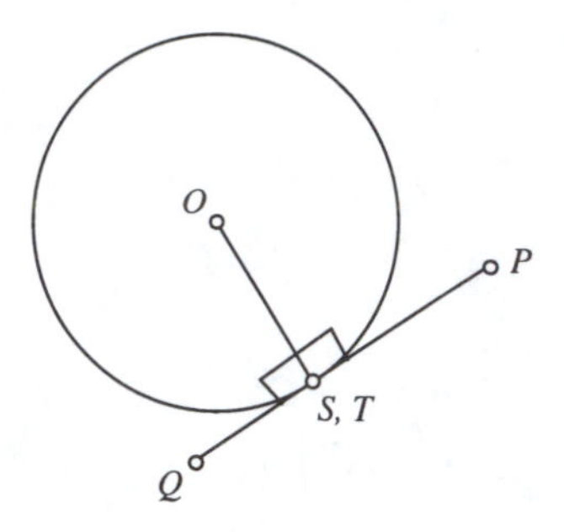

FIGURE 15

and in case (ii), we have the contradiction

$$PS + QT < PV + QU < PQ.$$

Hence S and T must be the same point, completing the proof.

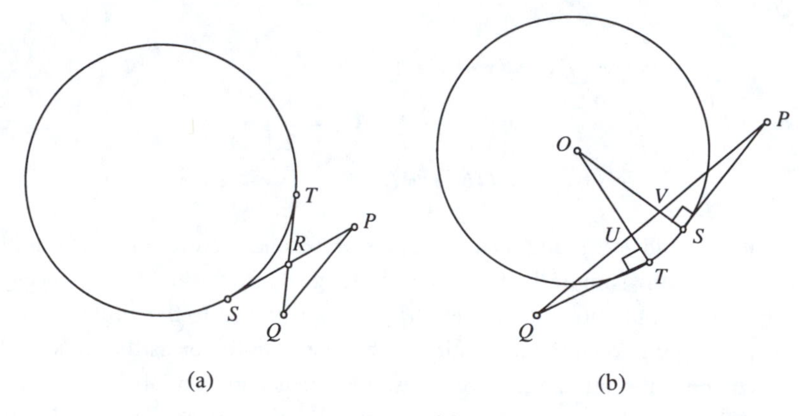

FIGURE 16

15. A and B are acute angles in $\triangle ABC$. If $\angle A = 30°$ and the altitude from A is the same length as the median from B, determine $\angle B$ (Figure 17).

Let MK be the perpendicular from M to BD (Figure 18). Since M is the midpoint of AC and MK is parallel to AD, then $MK = \frac{1}{2}AD$, and since $AD = MB$, then $MK = \frac{1}{2}MB$, implying MBK is a 30°-60°-90° triangle with $\angle MBK = 30°$.

With 30° angles at A and B, and $\angle C$ common, triangles ABC and MBC are similar. Let $\angle BMC = \angle ABC = x$. Then, recalling that $AD = BM$, we have

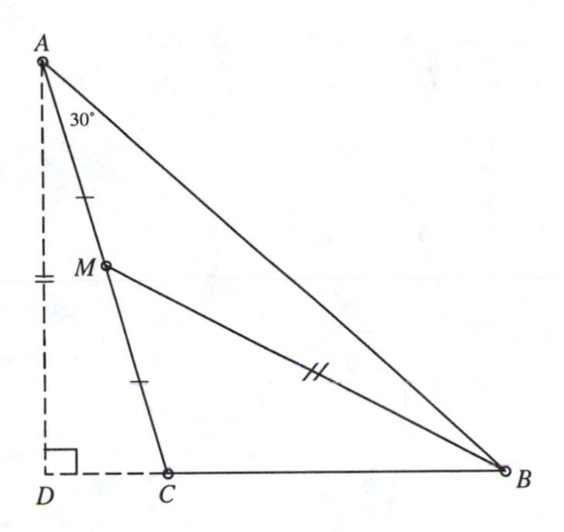

FIGURE 17

$$\sin \angle ABD = \sin x = \frac{AD}{AB} = \frac{BM}{AB} = \frac{\sin 30}{\sin \angle AMB} \qquad \text{(law of sines in } \triangle AMB).$$

Thus

$$\sin x = \frac{1/2}{\sin(180^\circ - \angle AMB)} = \frac{1/2}{\sin \angle BMC} = \frac{1/2}{\sin x}.$$

Hence

$$\sin^2 x = \frac{1}{2}, \quad \sin x = \frac{1}{\sqrt{2}},$$

and

$$\angle ABC = x = 45^\circ.$$

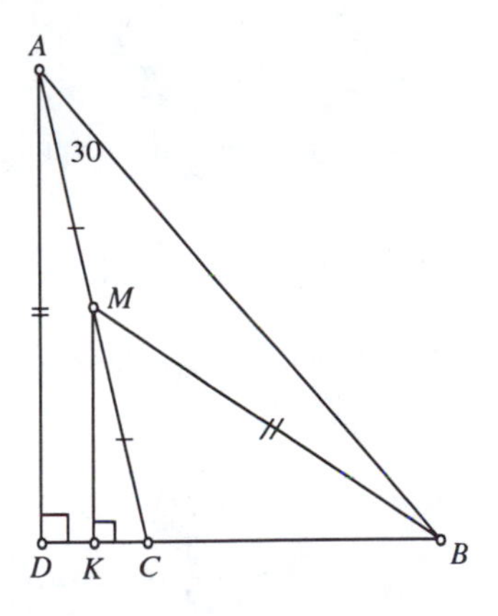

FIGURE 18

16. The incircle of $\triangle ABC$ touches BC at X. Prove that the line which joins the midpoints M and D of AX and BC passes through the incenter I (Figure 19).

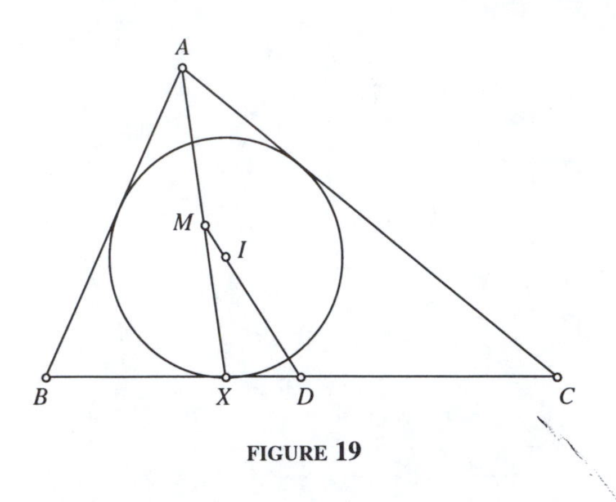

FIGURE 19

Let us prove the equivalent fact that DI extended bisects AX. We will use the theorem that the line from the midpoint of one side of a triangle, parallel to another side, bisects the third side. To this end, let Y be taken on BC so that $DY = DX$ (Figure 20), and let us show that DI and AY are parallel. This can be done by showing that the angles $\alpha = \angle XDI$ and $\beta = \angle DYA$ are equal; it is simply a matter of performing the calculations to show that $\tan\alpha = \tan\beta$.

If AE is the altitude to BC, the area Δ of triangle ABC is given by $\Delta = \frac{1}{2}a \cdot AE$, giving

$$AE = \frac{2\Delta}{a} = \frac{2rs}{a}$$

(since $\Delta = rs$, where r is the inradius and s the semiperimeter). Then

$$\tan\beta = \frac{AE}{EY} = \frac{2rs}{a \cdot EY}.$$

It is well known that the tangent BX has length $s - b$, and since $BE = c\cos B$, we obtain

$$\begin{aligned} EY = ED + DY = ED + DX &= (BD - BE) + (BD - BX) \\ &= \left(\frac{a}{2} - c\cos B\right) + \left(\frac{a}{2} - s + b\right) \\ &= a + b - s - c\cos B. \end{aligned}$$

Hence

$$\tan\beta = \frac{2rs}{a\cdot EY} = \frac{2rs}{a(a+b-s-c\cos B)}.$$

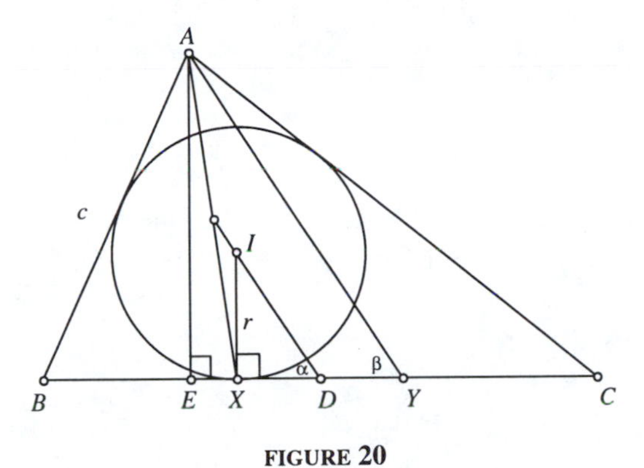

FIGURE 20

Now, from right triangle IXD, we have

$$\tan\alpha = \frac{IX}{XD} = \frac{r}{XD}.$$

Thus we want to show that

$$\frac{2rs}{a(a+b-s-c\cos B)} = \frac{r}{XD} = \frac{r}{\frac{a}{2}-s+b},$$

which reduces to

$$\frac{2s}{a(a+b-s-c\cos B)} = \frac{2}{a-2s+2b},$$

and, since $a-2s+2b = a-(a+b+c)+2b = b-c$, to

$$\frac{s}{a(a+b-s-c\cos B)} = \frac{1}{b-c},$$

giving

$$\left(\frac{a+b+c}{2}\right)(b-c) = a(a+b-s-c\cos B)$$

and

$$ab-ac+b^2-c^2 = 2a^2+2ab-2as-2ac\cos B.$$

Now, from the law of cosines, $2ac\cos B = a^2 + c^2 - b^2$, and so we want to show that

$$\begin{aligned} ab - ac + b^2 - c^2 &= 2a^2 + 2ab - 2as - a^2 - c^2 + b^2 \\ 0 &= a^2 + ab + ac - a(2s) \\ &= a(a+b+c) - a(a+b+c), \end{aligned}$$

which is clearly so.

17. Determine a point P inside $\triangle ABC$ such that the product of the distances from P to the three sides of the triangle is a maximum.

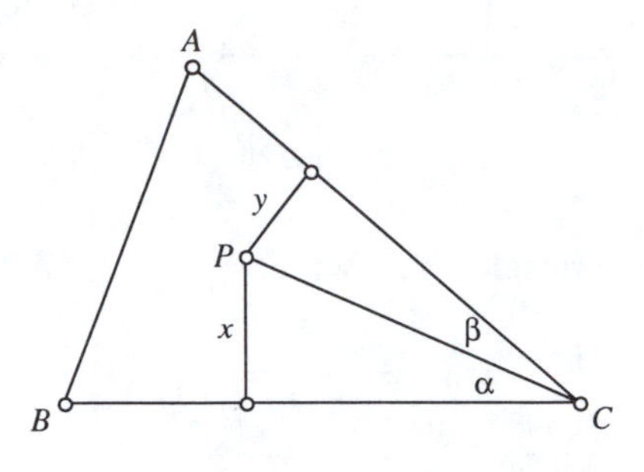

FIGURE 21

Suppose PC divides angle C into parts α and β (Figure 21). Then the perpendiculars x and y to the sides BC and AC are given by

$$x = PC\sin\alpha \quad \text{and} \quad y = PC\sin\beta,$$

and therefore

$$\begin{aligned} xy &= PC^2\sin\alpha\sin\beta \\ &= PC^2\left\{-\frac{1}{2}[\cos(\alpha+\beta) - \cos(\alpha-\beta)]\right\} \\ &= \frac{1}{2}PC^2[\cos(\alpha-\beta) - \cos C] \quad (\text{since } \alpha+\beta = C) \\ &\le \frac{1}{2}PC^2(1-\cos C) \\ &= \frac{1}{2}PC^2\left[1 - \left(1 - 2\sin^2\frac{C}{2}\right)\right] \end{aligned}$$

$$= \left(PC \sin \frac{C}{2} \right)^2$$

$$= \text{the product } xy \text{ when } P \text{ lies on the bisector of } \angle C.$$

Surely, then, the required point P is the incenter I, which lies on all three angle bisectors and has the perpendiculars $x = y = z = r$, the inradius. Fortunately, such a nice analysis is not always so misleading; as we shall see in the following beautiful solution, the condition that maximizes the product of two of the perpendiculars has no bearing on the condition which maximizes the full product xyz.

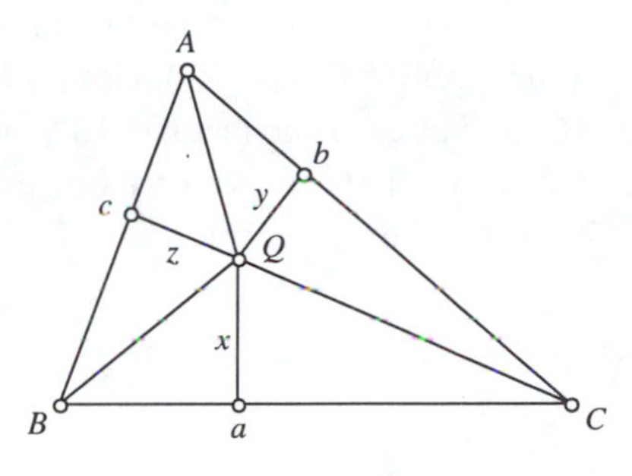

FIGURE 22

For any point Q inside $\triangle ABC$, the segments AQ, BQ, CQ partition $\triangle ABC$ into three smaller triangles whose areas are $\frac{1}{2}ax$, $\frac{1}{2}by$, $\frac{1}{2}cz$ (Figure 22). Doubling and adding, we get

$$ax + by + cz = 2\Delta, \quad \text{where } \Delta \text{ is the area of } \triangle ABC.$$

Now, the decisive step lies in thinking of the arithmetic mean–geometric mean inequality. From this useful result we have

the geometric mean $\leq$ the arithmetic mean,

and so

$$(ax \cdot by \cdot cz)^{1/3} \leq \frac{ax + by + cz}{3} = \frac{2\Delta}{3},$$

implying

$$abc \cdot xyz \leq \frac{8\Delta^3}{27}$$

and

$$xyz \leq \frac{8\Delta^3}{27abc},$$

with equality if and only if $ax = by = cz$.

That is to say, the maximum product is obtained only when AP, BP, CP divide $\triangle ABC$ into three triangles of equal area. It follows easily, then, that P must be the centroid G of $\triangle ABC$:

> Clearly, if a point lies on each of two straight lines, it must be their point of intersection. Since $\triangle PBC = \frac{1}{3}\triangle ABC$, then x must equal $\frac{1}{3}$ the altitude from A, and therefore P must lie on the line ST parallel to BC that trisects the altitude from A (Figure 23). Now, ST trisects every segment from A to BC, in particular the median AM, and so ST must pass through the centroid G. Similarly, P lies on a line parallel to AC which goes through G. Thus P can only be their point of intersection G.

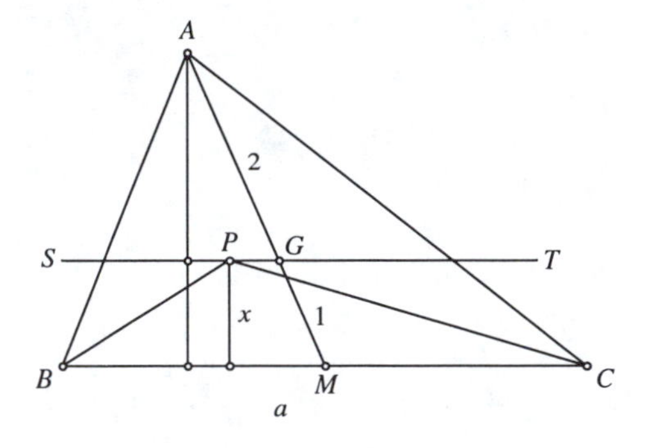

FIGURE 23

18. At an arbitrary point C on diameter AB of a circle with center O, a perpendicular is drawn to meet the circle at D (Figure 24). The circle inscribed in the curvilinear triangle DCB touches AB at J.
    (a) Prove that $AJ = AD$.
    (b) Prove that DJ bisects $\angle CDB$.

(a) Since it is not obvious how to construct the inscribed circle of DCB, and thus get a line on its radius, it could be that the problem is complicated. Thus it is a pleasant surprise to discover how really simple it is.

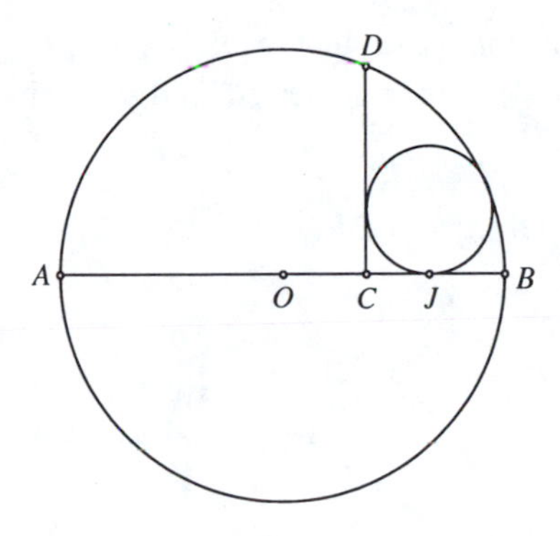

FIGURE 24

Let the center of the incircle of DCB be P and its radius r (Figure 25), and let the radius of the circle (O) be R. Since the circle (P) is internally tangent to (O), the distance OP between their centers is the difference of their radii, and in right triangle OPJ we have

$$OJ^2 = OP^2 - PJ^2 = (R-r)^2 - r^2 = R^2 - 2Rr.$$

Therefore

$$\begin{aligned} AJ^2 &= (AO + OJ)^2 = (R + OJ)^2 \\ &= R^2 + 2R \cdot OJ + OJ^2 \\ &= R^2 + 2R \cdot OJ + R^2 - 2Rr \\ &= 2R^2 + 2R(OJ - r) \\ &= 2R^2 + 2R \cdot OC \quad (\text{since } r = CJ). \end{aligned}$$

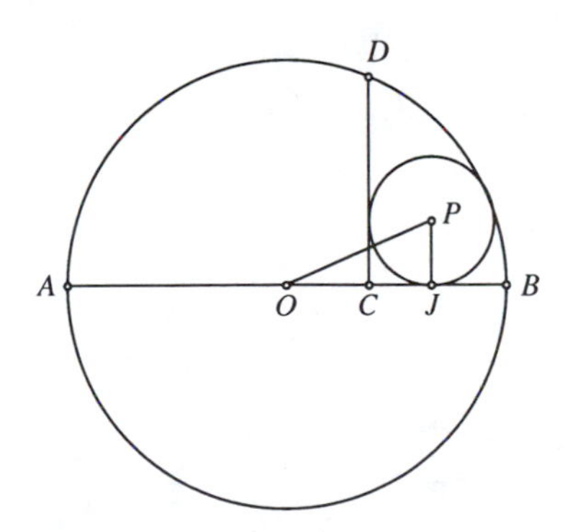

FIGURE 25

But it is almost immediate that AD^2 is given by the same expression. Since DC is the altitude to the hypotenuse in right triangle ADB, we have the mean proportion

$$AD^2 = AB \cdot AC = 2R(R + OC) = 2R^2 + 2R \cdot OC.$$

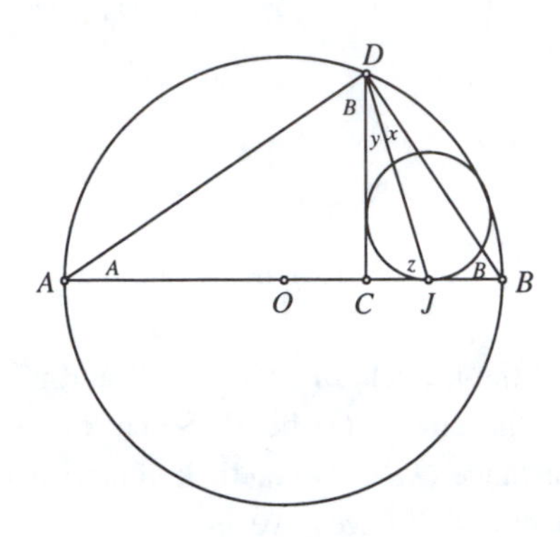

FIGURE 26

(b) In Figure 26, $\angle ADB$ is a right angle since AB is a diameter, and so $\angle ADC = 90° - (x + y) = \angle B$ in right triangle DCB.

Since $AD = AJ$, $\triangle ADJ$ is isosceles and $B + y = \angle DJA = z$. But z is an exterior angle of $\triangle JDB$, and $B + x = z$. Hence $x = y$, as desired.

19. In $\triangle ABC$, suppose D is in AB and E in AC such that $BD = DE = EC$ (Figure 27). If $\angle A = 60°$, prove that BE and CD intersect at the circumcenter O of $\triangle ABC$.

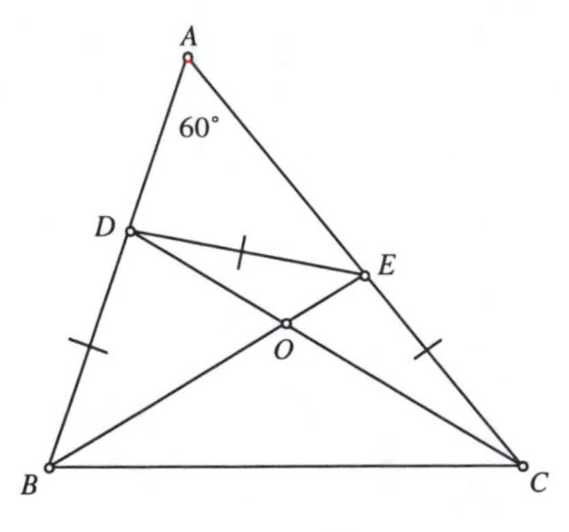

FIGURE 27

Suppose BE and DC intersect at P and that the equal base angles in isosceles triangles DBE and DEC are x and y (Figure 28); also, let $\angle PBC = s$ and $\angle PCB = t$. Hence exterior $\angle EPC$ of $\triangle DEP$ is equal to $x + y$. But $\angle EPC$ is also an exterior angle for $\triangle PBC$, and is equal to $s + t$. Thus, since

$$(x + y) + (s + t) = \angle B + \angle C,$$

we have

$$\begin{aligned} x + y &= s + t \\ &= \frac{1}{2}(\angle B + \angle C). \end{aligned}$$

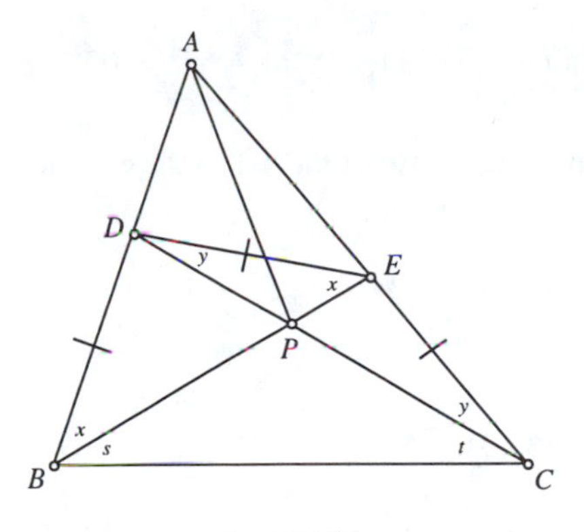

FIGURE 28

Since $\angle A = 60°$, then $\angle B + \angle C = 120°$, and we have

$$\begin{aligned} \angle EPC &= x + y \\ &= 60° = \angle A, \end{aligned}$$

making $ADPE$ a cyclic quadrilateral.

Hence $\angle DAP = \angle DEP = x$, making $\triangle ABP$ isosceles with $AP = BP$. Similarly, $\angle PAE = \angle PDE = y$, $\triangle PAC$ is isosceles and $AP = CP$. Thus $AP = BP = CP$, and P is indeed the circumcenter of $\triangle ABC$.

20. Evaluate

$$\int \frac{5}{16 + 9\cos^2 x}\, dx.$$

Beginning with the unusual move of substituting $16(\cos^2 x + \sin^2 x)$ for 16, we get

$$\int \frac{5}{16 + 9\cos^2 x}\,dx = 5\int \frac{dx}{25\cos^2 x + 16\sin^2 x},$$

and dividing numerator and denominator by $\cos^2 x$, we obtain

$$5\int \frac{\sec^2 x\,dx}{25 + 16\tan^2 x} = \frac{1}{5}\int \frac{d(\tan x)}{1 + \left(\frac{4}{5}\tan x\right)^2} = \frac{1}{4}\int \frac{d\left(\frac{4}{5}\tan x\right)}{1 + \left(\frac{4}{5}\tan x\right)^2}$$

$$= \frac{1}{4}\arctan\left(\frac{4}{5}\tan x\right) + C.$$

21. Solve the equation $3^{2x} - 34(15^{x-1}) + 5^{2x} = 0$.

Multiplying by 15 to get rid of the -1 in the exponent of the middle term, and writing 15^x as $3^x \cdot 5^x$, we get

$$15 \cdot 3^{2x} - 34 \cdot 3^x \cdot 5^x + 15 \cdot 5^{2x} = 0,$$

which factors into

$$(3 \cdot 3^x - 5 \cdot 5^x)(5 \cdot 3^x - 3 \cdot 5^x) = 0,$$

i.e.,

$$(3^{x+1} - 5^{x+1})(5 \cdot 3^x - 3 \cdot 5^x) = 0.$$

Hence

(i) $3^{x+1} - 5^{x+1} = 0$ gives $x + 1 = 0$, and $x = -1$,

and

(ii) $5 \cdot 3^x - 3 \cdot 5^x = 0$ gives $15(3^{x-1} - 5^{x-1}) = 0$,

$3^{x-1} - 5^{x-1} = 0$, $x - 1 = 0$ and $x = 1$.

Thus the roots are $x = \pm 1$.

22. Equilateral triangles PAB and QAC are drawn outwardly on the sides of $\triangle ABC$. Prove that the midpoint R of BC, the centroid G of $\triangle QAC$, and the vertex P always determine a 30°-60°-90° triangle (Figure 29).

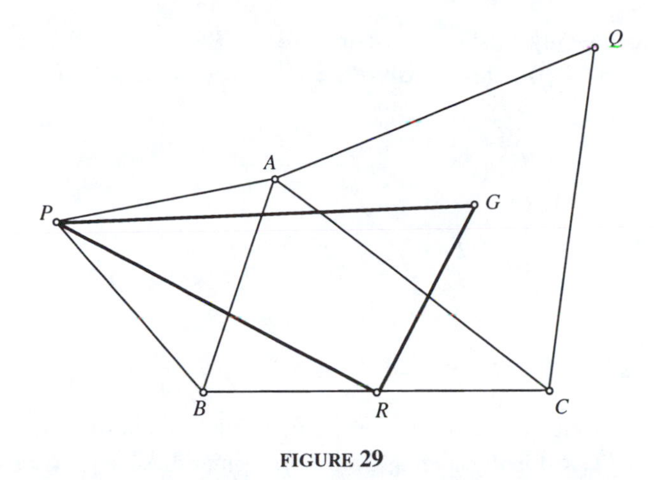

FIGURE 29

It suffices to show that $\angle PGR = 60°$ and that $PG = 2GR$.

To this end, let T be the midpoint of AC (Figure 30). Since R is the midpoint of BC, then RT is parallel to AB and half as long, and since $\triangle PAB$ is equilateral, then $PA = AB$ and we have $PA = 2RT$. This is the first step in showing that triangles PAG and RTG are similar with sides in the ratio 2 : 1. The proof of this is completed as follows.

Since the centroid G of an equilateral triangle is also its circumcenter, GT is perpendicular to AC. Also

$$\angle AGC = 2\angle Q = 120°, \quad \text{making} \quad \frac{1}{2}\angle AGC = \angle AGT = 60°,$$

and so $\triangle AGT$ is a 30°-60°-90° triangle. Hence $AG = 2GT$.

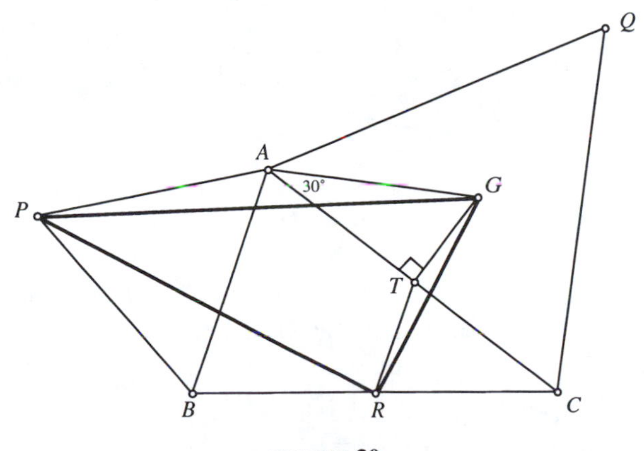

FIGURE 30

It remains only to show that our triangles PAG and RTG have equal angles at A and T. But this is easy: since TR is parallel to AB, then

$$\angle RTC = \angle BAC,$$

making

$$\angle RTG = \angle BAC + 90^\circ;$$

and at A,

$$\angle PAG = \angle PAB + \angle BAC + \angle TAG = \angle BAC + 90^\circ.$$

Thus we have the desired $PG = 2GR$, and the similar triangles also give $\angle AGP = \angle TGR$. Finally, adding $\angle PGT$ to each of these angles, we get

$$\angle PGR = \angle AGT = 60^\circ.$$

23. In equilateral triangle ABC, K is any circle which is contained entirely inside $\triangle ABC$ and has center at the center P of $\triangle ABC$. Tangents BR and CS are drawn as in Figure 31. Prove that SR bisects BC.

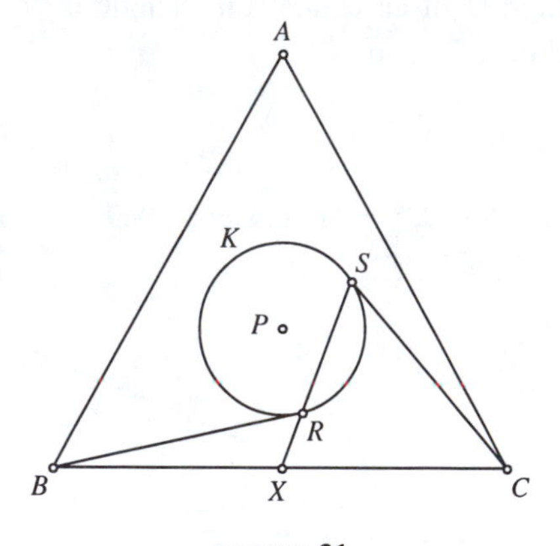

FIGURE 31

Let the midpoint of BC be M (Figure 32). Then PM is perpendicular to BC and BP subtends right angles at M and R, making $BPRM$ a cyclic quadrilateral. Also, $\angle BPC = 120^\circ$ and $\angle BPM = \frac{1}{2}\angle BPC = 60^\circ$, and therefore

$$\angle BRM = \angle BPM = 60^\circ \quad \text{(on chord } BM\text{)},$$

making

$$\angle PRM = 90^\circ + 60^\circ = 150^\circ \quad \text{and} \quad \angle PRY = 30^\circ.$$

That is to say, the line through R which bisects BC is inclined to PR at 30°. It remains, then, only to show that $\angle PRS = 30^\circ$.

Since $\angle BPC = 120^\circ$, the rotation about P through 120° carries B to C and tangent BR to tangent CS, implying $\angle RPS = 120^\circ$ in isosceles triangle PRS (Figure 31). Therefore base angle $\angle PRS = 30^\circ$, as desired.

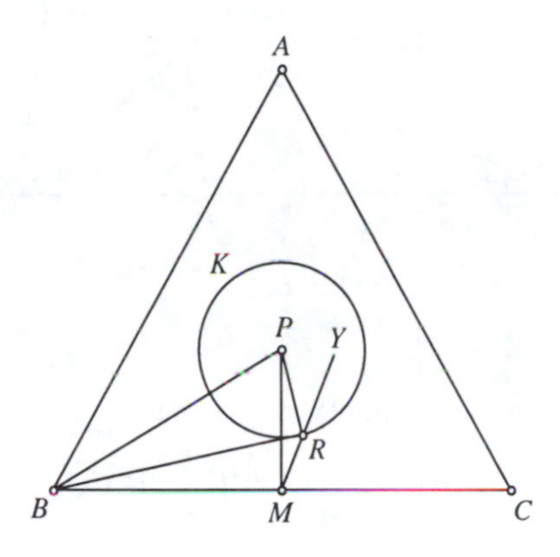

FIGURE 32

24. In isosceles triangle ABC the base angles at B and C are 40°. The bisector of angle B meets AC at D and BD is extended to E so that $DE = AD$ (Figure 33). How big is $\angle E$?

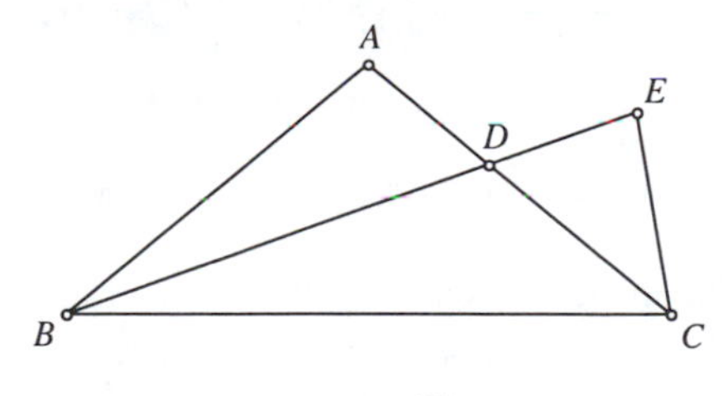

FIGURE 33

Since $\angle ADB$ is an exterior angle of $\triangle DBC$, we have

$$\angle ADB = \angle DBC + \angle DCB = 20^\circ + 40^\circ = 60^\circ.$$

Because BD bisects angle B, reflecting A in BD would take it to a point F on BC (Figure 34) and carry AD to DF and $\angle ADB$ to $\angle BDF$. Hence $FD = AD = DE$, and $\angle BDF = 60°$. Since ADC is a straight angle, $\angle FDC$ is also $60°$, and then, similarly, $\angle EDC$ is yet another $60°$ angle.

Hence $\triangle DFC \equiv \triangle DEC$ (SAS), and

$$\angle E = \angle DFC = 180° - 60° - 40° = 80°.$$

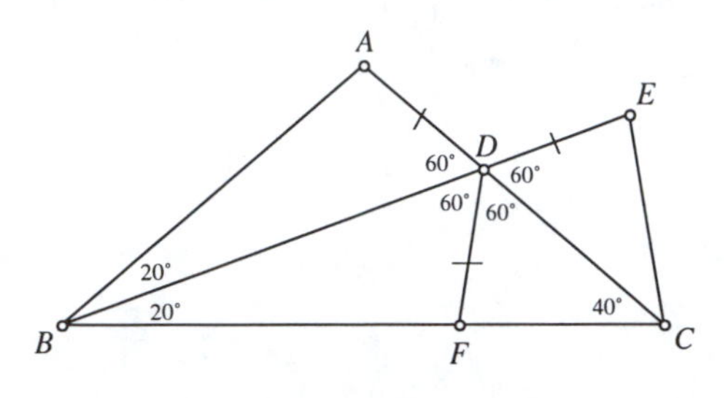

FIGURE 34

25. For positive real numbers x and y, prove that

$$(x + y)(xy + 1) \geq 4xy.$$

Solution 1 Multiplying out, the given inequality is

$$x^2y + xy^2 + x + y \geq 4xy.$$

Applying the arithmetic mean–geometric mean inequality directly to the four terms on the left side, we get the desired result immediately upon multiplying by 4:

$$\frac{x^2y + xy^2 + x + y}{4} \geq (x^4y^4)^{1/4} = xy.$$

In addition, we have that equality holds if and only if

$$x^2y = xy^2 = x = y, \quad \text{implying} \quad x^3 = x \quad \text{and} \quad x = y = 1.$$

Solution 2 Since the inequality contains the two factors $x + y$ and $xy + 1$, let us apply the A. M.–G. M. inequality to each factor separately. This gives

$$\frac{x + y}{2} \geq \sqrt{xy} \quad \text{and} \quad \frac{xy + 1}{2} \geq \sqrt{xy}.$$

Then

$$(x + y)(xy + 1) \geq (2\sqrt{xy})^2 = 4xy, \quad \text{as desired.}$$

Solution 3 Since x and y are positive, then clearly

$$x(y-1)^2 + y(x-1)^2 \geq 0,$$
$$x(y^2 - 2y + 1) + y(x^2 - 2x + 1) \geq 0,$$
$$xy^2 - 2xy + x + yx^2 - 2xy + y \geq 0,$$
$$x^2y + xy^2 + x + y \geq 4xy,$$

that is,

$$(x+y)(xy+1) \geq 4xy.$$

Solution 4 It is well known for $a > 0$ that $a + \frac{1}{a} \geq 2$. Hence

$$x + \frac{1}{x} + y + \frac{1}{y} \geq 4,$$

and clearing of fractions, we obtain

$$x^2y + y + xy^2 + x \geq 4xy, \quad \text{as desired.}$$

26. Prove that if each point of the plane is colored red, white, or blue, some unit segment will have both its endpoints the same color.

Consider two abutting equilateral triangles ABC and BCD of unit side (Figure 35). In order to avoid a monochromatic unit segment, the three vertices of each of these triangles must receive a different color, forcing the same color on A and D. Now, $AD = \sqrt{3}$, and it follows that, in order to avoid a monochromatic unit segment, every segment of length $\sqrt{3}$ must be monochromatic.

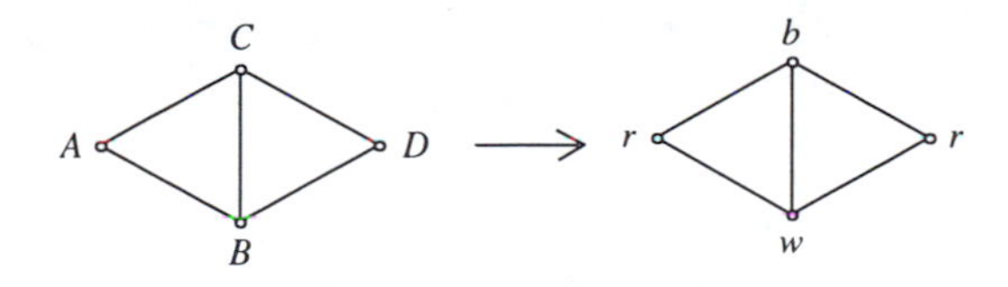

FIGURE 35

Consequently unless every circle K of radius $\sqrt{3}$ has all its points the same color as its center, a forbidden monochromatic unit segment must arise; and since points XY which are 1 unit apart on such a circle K themselves determine a monochromatic unit segment, the forbidden configuration is impossible to avoid (Figure 36(a)).

Alternatively, one might observe that the three vertices of an isosceles triangle with arms of length $\sqrt{3}$ must have all its vertices the same color. Thus every such triangle with unit base contains a monochromatic unit segment (Figure 36(b)).

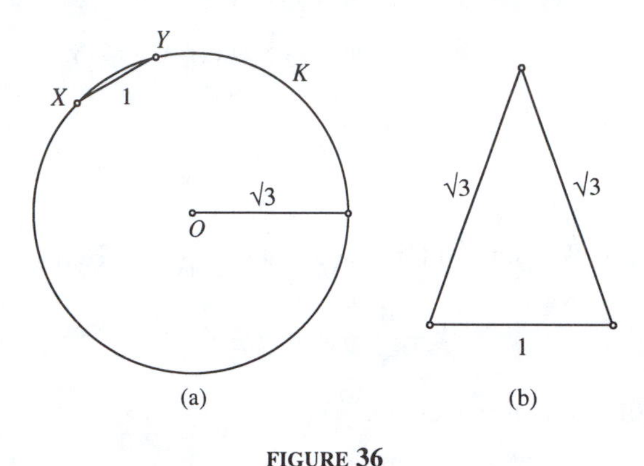

FIGURE 36

27. Find a graph having exactly three edges at each vertex whose edges cannot each be colored red, white, or blue, so that there is an edge of each color at each vertex.

In the graph G in Figure 37, let's try to make the edges at each of A, B, C, and D a different color. Since it is immaterial how the colors are distributed at the first vertex, let us begin at A as shown. Then BD would have to be blue in order to avoid either two red edges at B or two white ones at D. Thus at B, the third edge BC must be white, and at D the third edge DC must be red. Finally, at C, then, CE must be blue, providing a second blue edge at E.

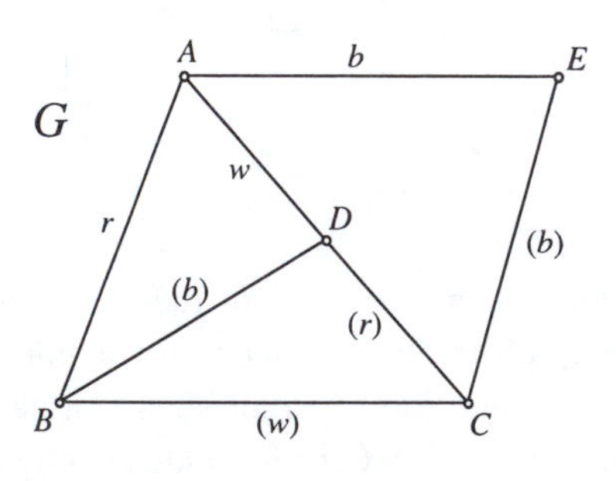

FIGURE 37

Thus G would constitute a solution to the problem if it weren't for E having only two edges. To complete the solution, then, we need to extend G into a graph which is cubic at every vertex. Clearly this is easily accomplished with a little reflection.

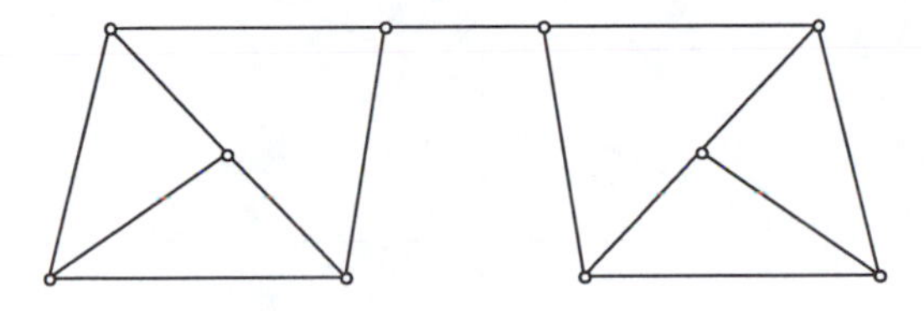

FIGURE 38

28. Given a circle C interior to an angle AOB (Figure 39), determine a straightedge and compass construction for the point P on C such that the sum of its distances to the arms of the angle is a minimum.

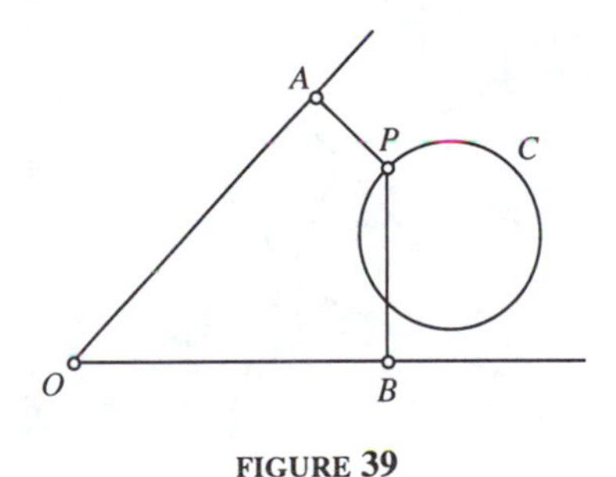

FIGURE 39

This an opportunity to use the fact that the sum $s+t$ of the distances to the arms of an angle XYZ from a point Q inside the angle (Figure 40) is given by the altitude LN of the isosceles triangle LYM which is cut from the angle by the line through Q perpendicular to the bisector YW of the angle: since $LY = YM$, we have

$$\triangle LYM = \triangle QYM + \triangle QYL = \frac{1}{2}YM \cdot s + \frac{1}{2}LY \cdot t$$
$$= \frac{1}{2}YM(s+t) = \frac{1}{2}YM \cdot LN, \Rightarrow s+t = LN.$$

(I first came across this marvellous result in the work of Murray Klamkin, University of Alberta.)

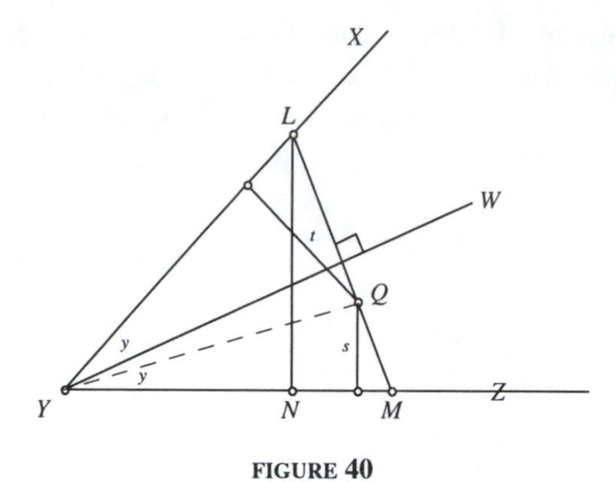

FIGURE 40

Thus the smallest triangle LYM yields the shortest altitude LN and the smallest sum $s+t$. Therefore the required point P is the point of contact of the tangent LM which is perpendicular to the bisector OW (Figure 41). Clearly P is determined by the line through the center K of C which is parallel to OW.

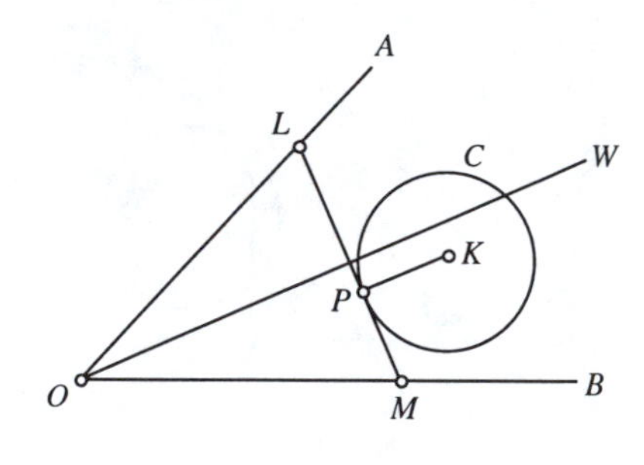

FIGURE 41

References

All references are to the Pi Mu Epsilon Journal.

1. 1950, **52**, Problem 2; proposed by the anonymous Problems Editor; solved by Lawrence Bennett, Brooklyn College.
2. 1951, **185**, Problem 16; proposed by W. J. Jenkins, Livingston, Alabama; solved by Mel Stover, Winnipeg, Manitoba.
3. 1951, **282**, Problem 44; proposed by Paul W. Gilbert, Syracuse University; solved by Leon Bankoff, Los Angeles, California.
4. 1951, **151**, Problem 24; proposed by Paul J. Schillo, University of Buffalo; solved by G. E. Raynor, Lehigh University.

5. 1952, **280**, Problem 38; proposed by C. W. Trigg, Los Angeles City College; solved by R. W. Hippe, St. Louis University.
6. 1951, **177**; from the article "On The Equation $\varphi(n) = \pi(n)$," by Leo Moser, University of Alberta.
7. 1951, **151**, Problem 23; proposed by Roy Dubisch, Fresno State College; solved by Howard Eves, Oregon State College.
8. 1954, **26**, Problem 63; proposed by Leon Bankoff, Los Angeles, California; solved by C. W. Trigg, Los Angeles City College.
9. 1958, **422**, Problem 97; proposed and solved (solution 2) by Alan J. Goldman, Princeton University; solution 1 by Jack R. Porter, Oklahoma University; solution 3 is of my own devising.
10. 1964, **20**, Problem 152; proposed by Leo Moser, University of Alberta; solved by Leonard Carlitz, Duke University.
11. 1967, **297**, Problem 183; proposed by R. Penney, Ford Scientific Laboratory; solved by Stan Rabinowitz, Polytechnic Institute of Brooklyn.
12. 1969, **24**; adapted from Problem 227, proposed by R. Sivaramkrishnan, Government Engineering College, Trichur, South India.
13. 1971, **203**, Problem 256; proposed by R. S. Luthar, University of Wisconsin, Janesville.
14. 1971, **242**, Problem 239; proposed anonymously; solved by Charles W. Trigg, San Diego, California.
15. 1976, **311**, Problem 351; proposed by Jack Garfunkel, Forest Hills High School, Flushing, New York; solved by Zelda Katz, Beverly Hills, California.
16. 1978, **483**, Problem 417; proposed by Clayton Dodge, University of Maine, Orono Maine.
17. 1979, **76**, Problem 405; proposed by Norman Schaumberger, Bronx Community College, Bronx, New York; solution by Walter Blumberg and Norman Schaumberger.
18. Part (a): 1981, **265**, Problem 494; proposed by Zelda Katz, Beverly Hills, California. Part (b): 1988, **534**, Problem 675: proposed by John H. Scott, Macalester College, St. Paul, Minnesota.
19. 1981, **279**, Problem 473; proposed by Jack Garfunkel, Forest Hills High School, Flushing, New York; solved by Zelda Katz, Beverly Hills, California.
20. 1981, **278**, Problem 472; proposed by R. S. Luthar, University of Wisconsin Center, Janesville; solution by Yuan-Whay Chu, Janesville, Wisconsin.
21. 1983, **553**, Problem 519; proposed by Charles W. Trigg, San Diego, California; solved by Peter John Dombrowsky, University of Texas, Austin.
22. 1984, **57**, Problem 553; proposed by Jack Garfunkel, Flushing , New York; solved by Leon Bankoff, Los Angeles, California.
23. 1986, **267**, Problem 622; proposed by Walter Blumberg, Coral Springs, Florida.
24. 1987, **485**, Problem 637; proposed by R. S. Luthar, University of Wisconsin Center, Janesville; solved by Harry Sedinger and Charles Diminnie, St. Bonaventure University, St. Bonaventure, New York.

25. 1989, **65**, Problem 687; proposed by Basil Rennie, South Australia; solution 1 by Bob Prielipp, University of Wisconsin, Oshkosh; solution 2 by George P. Evanovich, St. Peter's College, Jersey City, New Jersey; solution 3 by Seung-Jin Bang, Seoul, Korea; solution 4 by the St. Olaf Problem Solving Class, St. Olaf College, Northfield, Minnesota.
26. 1984, **668**, Puzzle 6; proposed by Joseph Konhauser, Macalester College, St. Paul, Minnesota.
27. 1983, **608**, Puzzle 5; proposed by Joseph Konhauser, Macalester College, St. Paul, Minnesota.
28. 1986, **264**; adapted from Puzzle 5, proposed by Joseph Konhauser, Macalester College, St. Paul, Minnesota.

SECTION 7

Problems from the Austrian-Polish Mathematics Competitions

In 1994 the distinguished Polish mathematician Marcin E. Kuczma (University of Warsaw) published an English translation of the 144 problems that have appeared on the Austrian-Polish Mathematics Competitions from 1978 to 1993 (published by The Academic Distribution Center, 1216 Walker Rd., Freeland, Maryland, 21053, U. S. A.). One of his main purposes is to give the reader an opportunity to enjoy an individual involvement with the problems. Solving a problem at this level is a gratifying personal achievement. A great deal of thought has gone into the problems and the detailed solutions, and the volume is most warmly recommended.

The following problems are a small sample of its treasures. Problems from many subject areas and of varying degrees of difficulty are to be found in this collection. We begin with alternative solutions to the first four of the problems below; the remaining solutions are from the book, and it is a pleasure to thank Dr. Kuczma for his kind permission to discuss them here.

1. (1978)

Let $c \neq 1$ be a positive rational number. Prove that the positive integers can be partitioned into two disjoint subsets A and B so that the ratio of two numbers from the same subset is never equal to c, that is,

$$\text{for } x, y \in A, \ \frac{x}{y} \neq c; \quad \text{similarly for } x, y \in B.$$

Let $c = \frac{a}{b}$ in its lowest terms. Since $c \neq 1$, not both a and b can be 1; for definiteness, suppose $b \neq 1$.

Now, a positive integer n is uniquely expressible in the form $n = qb^r$, where b does not divide q, and $r \geq 0$; simply keep dividing n by b until it is no longer a factor: for example,

$$\text{for } b = 6, \quad 108 = 18 \cdot 6 = 3 \cdot 6^2, \quad \text{and} \quad 20 = 20 \cdot 6^0.$$

Then into the subset A put all the integers for which the exponent r is even and into B those with r odd.

Now suppose, for some $x, y \in A$, that

$$\frac{x}{y} = c = \frac{a}{b}, \quad \text{that is, } bx = ay.$$

Suppose x and y are given by the expressions

$$x = sb^{2t}, \quad \text{where } b \text{ does not divide } s,$$

and

$$y = ub^{2v}, \quad \text{where } b \text{ does not divide } u.$$

Then the integer n which is given by bx and ay is equal to both

$$sb^{2t+1} \quad \text{and} \quad aub^{2v}.$$

Since b does not divide u, and a and b are relatively prime, b does not divide au, implying that aub^{2v} is a second expression for n in the form qb^r. But these expressions cannot be the same, for one has an odd exponent of b and the other even. Thus no two integers in A can have a ratio equal to c without violating the uniqueness of this form qb^r. Similarly for the subset B, and the conclusion follows.

2. (1978)

   Consider 1978 sets, each containing 40 elements. If each two of the sets intersect in exactly one element, prove that there is an element that is common to all 1978 of the sets.

Let the sets be $A_1, A_2, \ldots, A_{1978}$. Since A_1 has a single common element with each of the other 1977 sets, some element t of the 40 elements of A_1 must assume this role at least 40 times $(1977 > 39 \cdot 40)$; for definiteness, suppose t is the intersection of A_1 with each of the 40 sets $A_2, A_3, \ldots, A_{41}$.

Let A be any other of the sets and suppose that t does not belong to A. Now, A has a single common element with each of $A_1, A_2, \ldots, A_{41}$, but no element s of A could be its common element with two of these sets A_m and A_n without each of A_m and A_n containing s in addition to t, in violation of the requirement that their intersection be a single element. But, containing only 40 elements, there is no way A can share a **different** element with each of the 41 sets $A_1, A_2, \ldots, A_{41}$. Thus if A does not contain the element t, the required conditions cannot be satisfied. Hence each of the other 1937 sets A must also contain t, and the conclusion follows.

3. (1984)

If each altitude of a tetrahedron meets the opposite face at its incenter, prove that the tetrahedron is regular.

Let the tetrahedron be $ABCD$, and let I be the incenter of $\triangle ABC$, r its inradius, and suppose that the incircle touches AB at E and AC at F (Figure 1). Then the altitude DI makes each of the triangles DIE and DIF right angled at I, and since they also have DI common and equal sides IE and IF ($= r$), the triangles are congruent. Hence $DE = DF$, making the three sides of $\triangle DAE$ respectively equal to the sides of $\triangle DAF$:

$DE = DF$, $\quad DA$ is common, and $AE = AF$ (equal tangents to the incircle).

Thus these triangles are also congruent and we have the important result that

$$\angle DAE = \angle DAF.$$

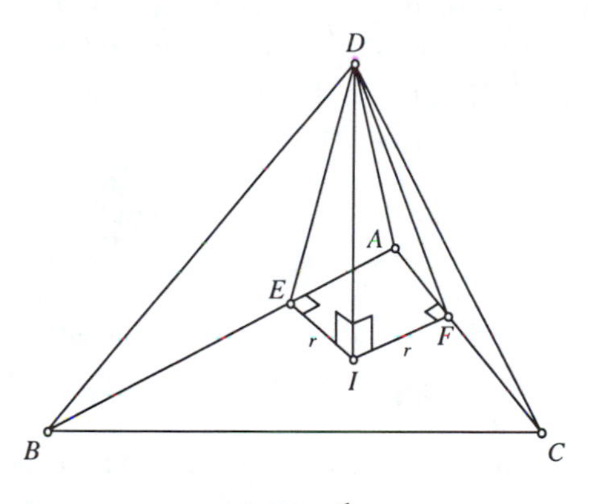

FIGURE 1

That is to say, two of the face angles at A are equal. Since a similar argument shows that the third face angle at A also belongs to a similar pair of equal angles, it follows that the three face angles at A must be equal. Similarly at the other vertices.

Let the equal face angles at A, B, C, D each be a, b, c, d, respectively. Then, adding up the angles in the four faces, we get

$$3a + 3b + 3c + 3d = 4 \cdot 180°,$$

which reduces to

$$a + b + c + d = 240°.$$

But the angles in the face ABC add up to 180°, and so

$$a + b + c = 180°,$$

implying

$$d = 60°.$$

Similarly, each of

$$a = b = c = d = 60°,$$

and the faces are all equilateral triangles, all the same size since each two of them share a common side. Hence $ABCD$ is regular.

4. (1985)

In a convex quadrilateral $ABCD$ of area 1, prove that the sum of the lengths of the sides and the diagonals is never less than $4 + \sqrt{8}$.

The isoperimetric theorem for plane polygons is quite well known and I expect it would be permissible to use it on this contest without proof (which is not an easy undertaking; a full proof is given in Russell Benson's *Euclidean Geometry and Convexity*, McGraw-Hill, 1966, pages 127–128):

of all plane n-gons with a given area, the one of minimum perimeter is regular.

Thus the perimeter of $ABCD$ is at least 4, the perimeter of a unit square. It remains to show that the sum of the diagonals is always at least $\sqrt{8}$. But this is easy.

Through A and C draw lines parallel to diagonal BD and through B and D draw lines parallel to the other diagonal AC to obtain parallelogram $PQRS$ (Figure 2). Then the perimeter of $PQRS$ is clearly twice the sum of the diagonals AC and BD and it is easy to see that its area is $2 \cdot ABCD = 2$:

$$\triangle ABD = \frac{1}{2} \text{ parallelogram } PBDS$$

and

$$\triangle CBD = \frac{1}{2} \text{ parallelogram } BQRD;$$

adding gives

$$ABCD = \frac{1}{2} PQRS,$$

and since $ABCD = 1$, then $PQRS = 2$.

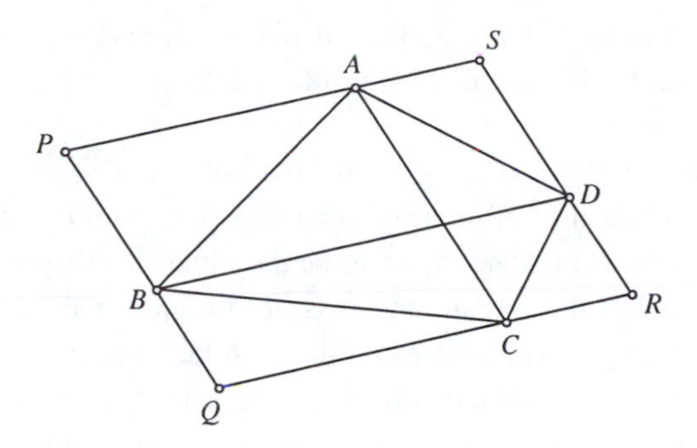

FIGURE 2

Since the area of $PQRS$ is always 2, its perimeter is at least that of a square of area 2, that is, of a square of side $\sqrt{2}$. Thus

$$2(AC + BD) \geq 4\sqrt{2},$$

giving the desired

$$AC + BD \geq 2\sqrt{2} = \sqrt{8}.$$

5. (1978)

F is a family of discs in the plane. Two discs may touch but there is no overlapping of interiors; in fact, each member is tangent to at least six other discs in the family. Prove that F is infinite.

Proceeding indirectly, suppose F is finite. In that case, some member A of F has minimum radius r. Being smallest, the six discs which touch A must each be as big as A. However, the tangents from the center O of A to a tangent disc of equal size subtends a 60° angle at O (Figure 3):

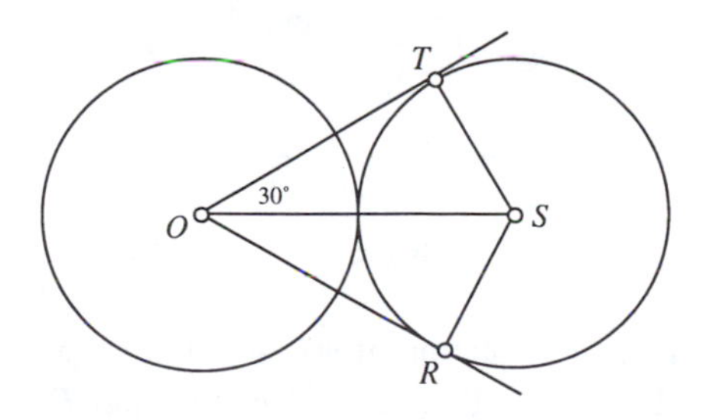

FIGURE 3

$\triangle OTS$ is right angled at T and OS is twice TS, making it a 30°-60°-90° triangle with the 30° angle at O; similarly $\angle ROS = 30°$.

Therefore no disc in the 360°-ring around A could be bigger than A without squeezing some other disc of the ring to a radius less than r. It follows, then, that A and its six touching discs must all be the same size (Figure 4).

Thus a disc B in the ring around A is of minimum radius r and accordingly has its own ring of six discs of radius r. In fact, the ring around B contains A and also a disc C which is directly opposite A. Similarly C, in turn, is touched by another disc D of radius r which is in line with A, B, and C. Clearly there is no end to this straight line of abutting discs of radius r and the conclusion follows by contradiction.

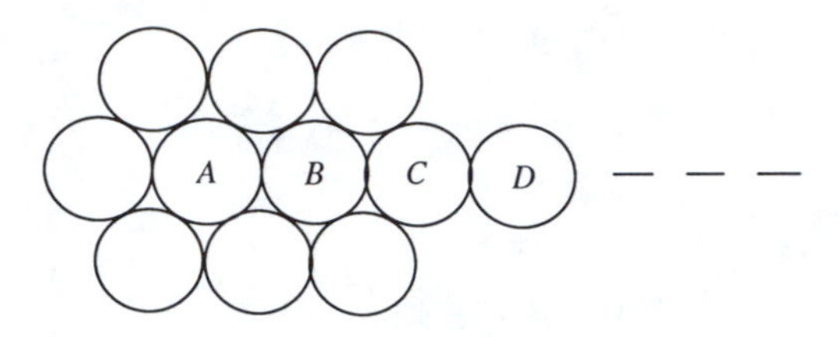

FIGURE 4

6. (1979)

E and F are points respectively on sides AB and BC of square $ABCD$ so that $BE = BF$ (Figure 5). If BN is an altitude of $\triangle BEC$, prove that $\angle DNF$ is a right angle.

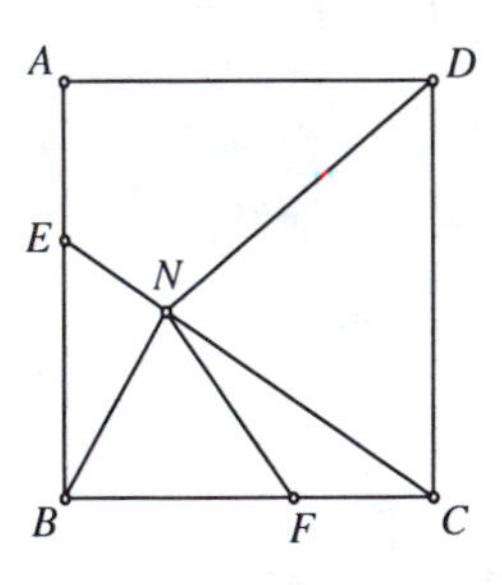

FIGURE 5

In Figure 6, a clockwise rotation of 90° about the center O of the square carries $\triangle BEC$ to $\triangle AMB$ such that M lies on AD, and EC is perpendicular to its image MB. Since BN is perpendicular to EC, N must lie on BM, and we have $\angle MNC = 90°$. Also, $AM = BE = BF$, making $MD = FC$ and $MFCD$ a rectangle. Thus the circle with diameter MC passes through the four vertices of the rectangle, and since $\angle MNC = 90°$, it also passes through N. But FD is another diameter of this circle, and therefore it also subtends a right angle at N.

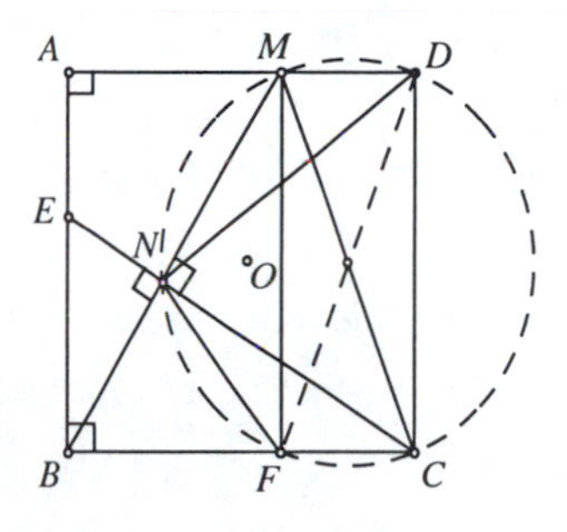

FIGURE 6

7. (1983)

Nonnegative real numbers a, b, x, and y are such that

$$a^5 + b^5 \le 1 \quad \text{and} \quad x^5 + y^5 \le 1.$$

Prove that also

$$a^2x^3 + b^2y^3 \le 1.$$

This gem is based on the brilliant observation that

$$a^2x^3 = \sqrt[5]{a^5a^5x^5x^5x^5}.$$

By the arithmetic mean–geometric mean inequality, then, we have

$$\frac{a^5 + a^5 + x^5 + x^5 + x^5}{5} \ge \sqrt[5]{a^5a^5x^5x^5x^5} = a^2x^3,$$

that is,

$$a^2x^3 \le \frac{2}{5}a^5 + \frac{3}{5}x^5.$$

Similarly,

$$b^2y^3 \le \frac{2}{5}b^5 + \frac{3}{5}y^5.$$

Hence, adding yields

$$a^2x^3 + b^2y^3 \le \frac{2}{5}(a^5 + b^5) + \frac{3}{5}(x^5 + y^5)$$
$$\le \frac{2}{5} \cdot 1 + \frac{3}{5} \cdot 1 = 1.$$

8. (1986)

If each point in space is colored red or blue, prove that

(i) all the vertices of some unit square are blue, or

(ii) none of the vertices of some unit square is blue, or

(iii) exactly one vertex of some unit square is blue.

If all the points in space are blue, the claim holds trivially.

Suppose, then, that some point O is red, and let S be the unit sphere with center O. If S is all blue, the claim again follows immediately since there are any number of unit squares $ABCD$ with four vertices on S (Figure 7; the vertices lie on the surface and the sides run through the interior).

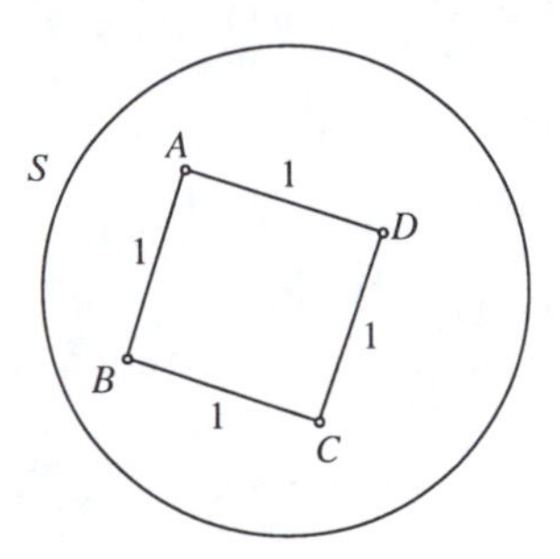

FIGURE 7

On the other hand, suppose some point P on S is red. Now forget S and construct a unit square $OPQR$ on the unit segment OP, and let it be spun around the axis OP through a complete revolution (Figure 8). The vertices Q and R thus describe two unit circles in parallel planes that are one unit apart.

If any point U of either of these circles is red, the corresponding unit square $OUVP$ has either no blue vertex or exactly one blue vertex, depending on the color of V. Otherwise, both circles are all blue and any unit chord KL gives rise to an all blue unit square $KLMN$ (Figure 9).

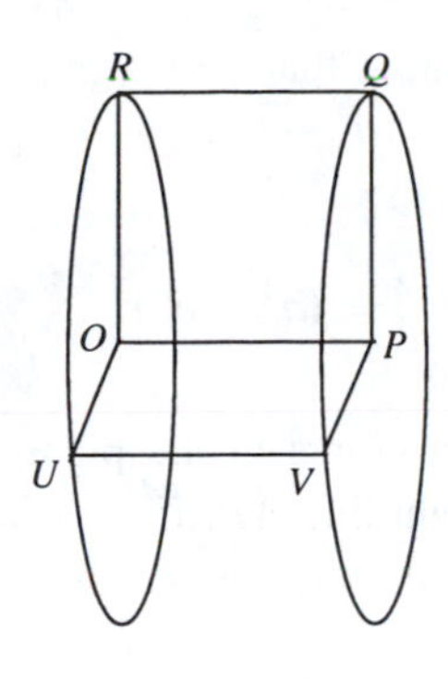

FIGURE 8

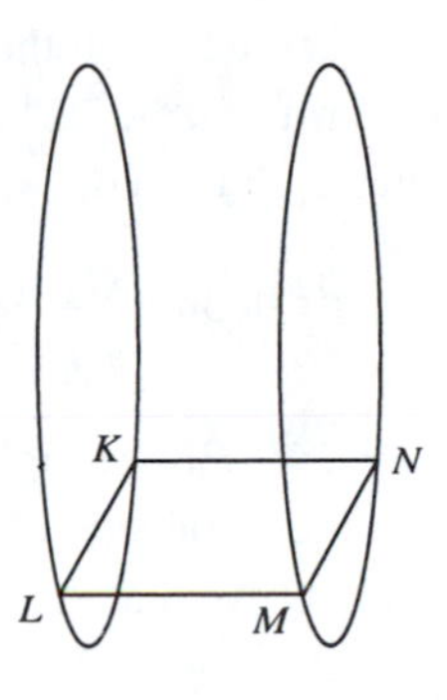

FIGURE 9

9. (1993)

Recall that the Fibonacci sequence is defined by

$$f_1 = 1, \; f_2 = 1, \quad \text{and for } n \geq 1, \quad f_{n+2} = f_{n+1} + f_n.$$

Now suppose that a and b are positive integers such that

$$b^{93} \text{ is divisible by } a^{19} \quad \text{and} \quad a^{93} \text{ is divisible by } b^{19}.$$

Prove that

$$(a^4 + b^8)^{f_{n+1}} \text{ is divisible by } (ab)^{f_n} \text{ for all } n > 1.$$

In one sense it's a pity this lovely-sounding problem turns out to be so easy.

It follows immediately that the claim holds for $n = k$ if it holds for the two previous values $n = k - 1$ and $n = k - 2$: if, for some positive integers t and s,

$$(a^4 + b^8)^{f_{k-1}} = t(ab)^{f_{k-2}}$$

and

$$(a^4 + b^8)^{f_k} = s(ab)^{f_{k-1}},$$

then, since $f_{n-2} + f_{n-1} = f_n$, multiplication gives

$$(a^4 + b^8)^{f_{k+1}} = ts(ab)^{f_k}.$$

In order to complete a solution by induction, then, it remains only to show that the hypothesis is valid for $n = 2$ and 3.

Since $a^{19}|b^{93}$, then $a^{19}|b^{95}$, which could not be true unless $a|b^5$. Thus $ab|b^6$, and similarly $ab|a^6$, which implies that $(ab)^2$ divides each of a^{12}, a^6b^6, and b^{12}, and we conclude with the observations that $n = 2$ gives $f_2 = 1$, and ab divides each term of $a^8 + 2a^4b^8 + b^{16} = (a^4 + b^8)^2$; $n = 3$ gives $f_3 = 2$, and $(ab)^2$ divides each term of

$$a^{12} + 3a^8b^8 + 3a^4b^{16} + b^{24} = (a^4 + b^8)^3.$$

10. $A, B, P_1, P_2, \ldots, P_6$ are eight distinct points in the plane with all the points P_i on the same side of the straight line AB. If the triangles P_iAB are all similar, prove that the six points P_i must lie on a circle.

Because similar triangles have the same trio of angles, it is the same trio (α, β, γ) that occurs in each triangle P_iAB. If two of these triangles were to have the same angle at A and also to have equal angles at B, their third vertices P_i and P_k would be the same point, a contradiction (Figure 10).

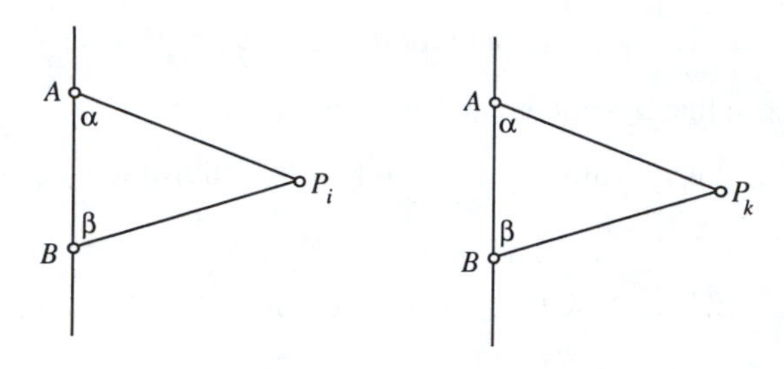

FIGURE 10

Therefore the ordered pair of angles at A and B must be different for each of the six triangles. But a trio (α, β, γ) gives rise only to six ordered pairs,

$$\{(\alpha, \beta), (\beta, \alpha), (\alpha, \gamma), (\gamma, \alpha), (\beta, \gamma), (\gamma, \beta)\},$$

and these are all different only when α, β, and γ are three different angles. Thus the triangles P_iAB are easily constructed by drawing each of α, β, and γ at each of A and B and following the arms of the pairs $(\alpha, \beta), (\beta, \alpha), \ldots,$ to the points P_i (Figure 11).

Now, the pairs (α, β) and (β, α) determine congruent triangles, each of which is obtained from the other by turning it over so as to interchange A and B, that is to say, by reflection in the perpendicular bisector XY of AB. The segment P_iP_k joining their third vertices is therefore parallel to AB and its perpendicular bisector is also XY; in Figure 11 the three such segments are labelled P_1P_2, P_3P_4, P_5P_6.

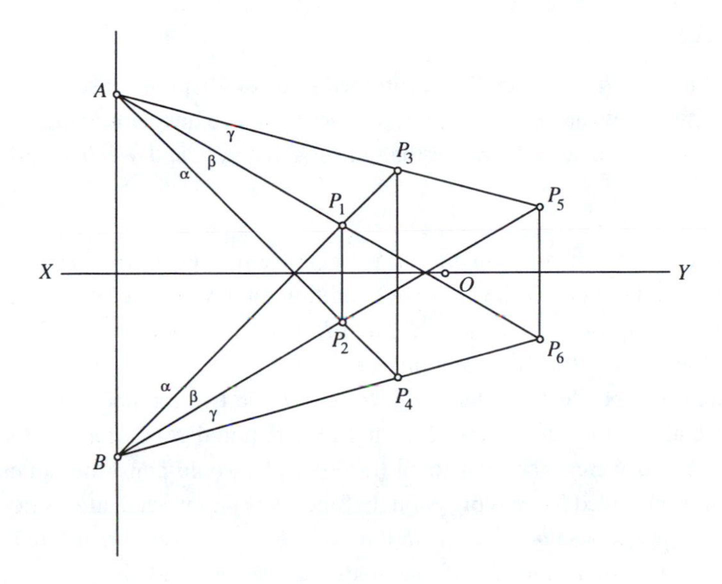

FIGURE 11

Now consider the quadrilateral $P_1P_2P_4P_6$. Since P_1P_2 is parallel to AB, the alternate angles at A and P_2 are equal:

$$\angle BAP_2 = \alpha = \angle AP_2P_1.$$

But in $\triangle ABP_6$, the angles at A and B are β and γ, and so the angle at P_6 must be α. Therefore

$$\angle AP_2P_1 = \angle P_4P_6P_1,$$

that is, the exterior angle of the quadrilateral $P_2P_4P_6P_1$ at P_2 is equal to the interior angle at the opposite vertex P_6, and we conclude that the quadrilateral is cyclic.

Now, any circle through P_1 and P_2 must have its center O on the perpendicular bisector XY of P_1P_2. Hence

$$OP_1 = OP_2 = OP_4 = OP_6.$$

But XY is also the perpendicular bisector of P_3P_4, and so

$$OP_3 = OP_4;$$

similarly,

$$OP_5 = OP_6,$$

and O is equidistant from all six points P_i, establishing their concyclic character.

11. (1983)

Let the positive integers be partitioned into two disjoint subsets A and B in any way whatever. Prove that, for every positive integer n, there are two integers a and b, each greater than n, such that a, b, and $a+b$ are all in A or all in B.

If A is finite, then B contains all the positive integers beyond some point and thus is able to satisfy the required condition for any value of n; similarly if B is finite. Suppose, then, that both A and B are infinite. This means that each subset contains arbitrarily large numbers.

Let n be specified and let a and b, $a < b$, be two members of A, each greater than n. Now, unless $a+b$ is in B, the required condition is satisifed. Also, if $b-a$ were to be in A, then $\{a, b-a, b\}$ would constitute a desired triple in A, provided $b-a$ is big enough. Since A is unbounded, after selecting $a > n$, simply choose $b > a+n$. In this case, $b-a > n$ and therefore $b-a$ would also have to reside in B if the condition is to be avoided.

$$A : \{\ldots, a, b, 2a, \ldots\},$$
$$B : \{\ldots, a+b, b-a, 3a, \ldots\}.$$

Since

$$(b-a) + 2a = a+b,$$

if one is to continue to avoid satisfying the hypothesis, there is now no option but to put $2a$ in A. Continuing to resist the desired condition, B must then hold $3a$. But in this case, $2a+b$ can't go in either A or B, for A already has both $2a$ and b, and B already has $3a$ and $(b-a)$. Thus the property in question is unavoidable.

SECTION 8
19 Problems from *Quantum*

(Specific references are given at the end.)

1. From an arbitrary point inside an equilateral triangle, segments to the vertices and perpendiculars to the sides partition the triangle into six little triangles, A, B, C, D, E, F (Figure 1). Prove that the areas

$$A + C + E = B + D + F.$$

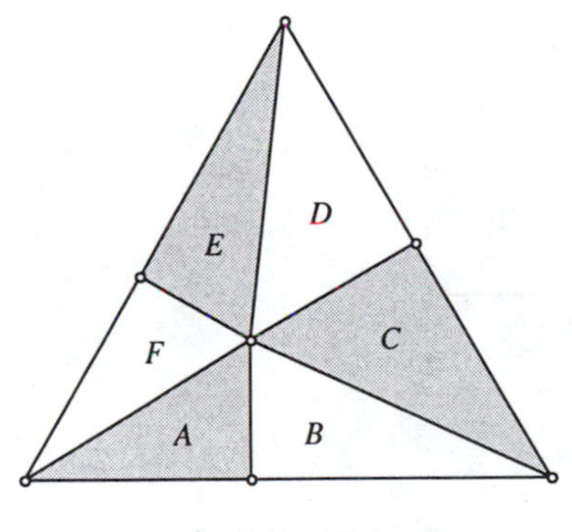

FIGURE 1

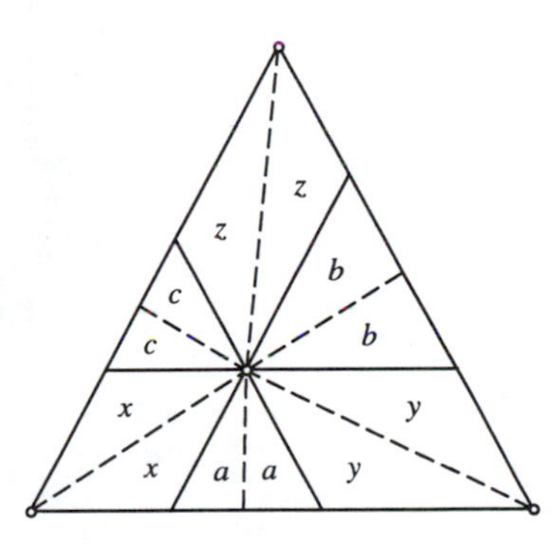

FIGURE 2

Clearly, three additional lines through the selected point which are parallel to the sides of triangle partition the triangle into three parallelograms and three small equilateral triangles (Figure 2). Since the areas of the parallelograms are bisected by their diagonals and the equilateral triangles by their altitudes, we have immediately that

$$A + C + E = x + a + y + b + z + c = B + D + F.$$

2. Forty-one rooks are placed arbitrarily on squares of a 10×10 chessboard. Prove that some five of the rooks determine a subset in which no two of them attack each other.

Suppose the board is rolled into a cyclinder (Figure 3). Now, for each of the 10 squares in the top row, there is a 10-step spiral staircase that winds around the cylinder exactly once while going from top to bottom. These 10 staircases cover the whole board and therefore contain the squares holding the 41 rooks. But 41 rooks cannot be spread over 10 staircases without some staircase getting 5 or more of them ($10 \cdot 4$ is only 40). Since no two squares of a staircase lie in the same row or the same column of the board, two rooks in the same staircase cannot attack one another, and so any staircase with 5 or more rooks yields a desired subset.

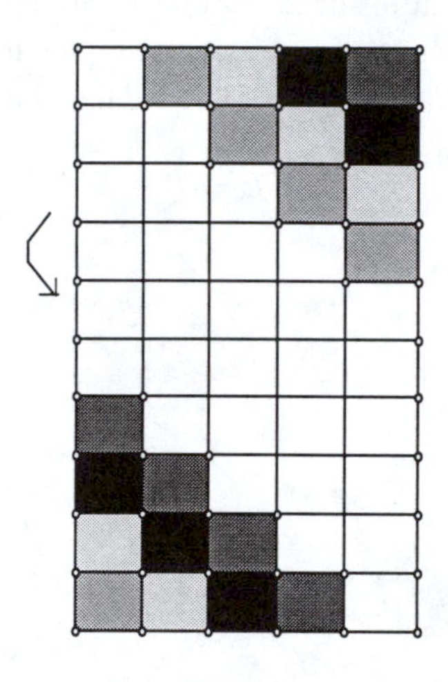

FIGURE 3

3. The product of a billion positive integers is equal to a billion. What is the greatest sum these billion numbers can have?

If two of the numbers a and b are both greater than 1, then

$$(a-1)(b-1) > 0,$$
$$ab - a - b + 1 > 0,$$

and

$$ab + 1 > a + b.$$

That is to say, the numbers ab and 1 yield a sum greater than $a + b$ without altering the product ($ab \cdot 1 = a \cdot b$). Therefore, changing all such pairs (a, b) to $(ab, 1)$, the maximum sum is obtained when all the numbers are 1 except the final ab. Since the total product is a billion, the final ab must itself be a billion, giving a maximum sum of (2 billion − 1).

4. S is a finite set of points in the plane, consisting of at least three points, not all collinear. Prove that some circle through three points of S has no points of S inside it.

One's first thought might be that the smallest circle through three points of S can't have any points of S inside it, but this is not so. If S consists of the vertices of an equilateral triangle and its center, the four possible circles through three points of S are all the same size, and so the circumcircle of the triangle is of minimum size but it does contain the center (Figure 4). Thinking small, however, is the key to a solution.

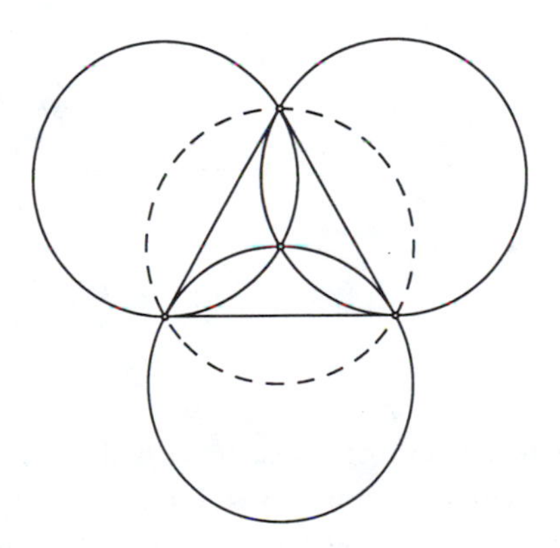

FIGURE 4

Suppose A and B are points of S which are at a minimum *distance* apart. Then the circle on diameter AB can't have any other point C of S inside or on it without C being closer to A or B than the diameter AB, in violation of the minimality of the distance AB. Now consider all the circles ABX where X runs through all the other points of S. If the circle ABP is the smallest of these, it can't have any other point R of S inside it without violating its own minimality, and the conclusion follows (Figure 5).

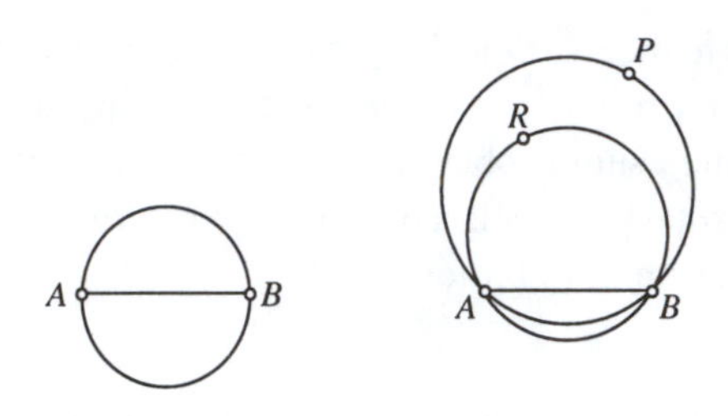

FIGURE 5

5. L is a finite set of straight lines in the plane, consisting of at least three lines, no two parallel and no three concurrent. Obviously these lines divide the plane into various regions. Prove that, for each line m in L, one of the regions adjacent to it must be a **triangle** (Figure 6).

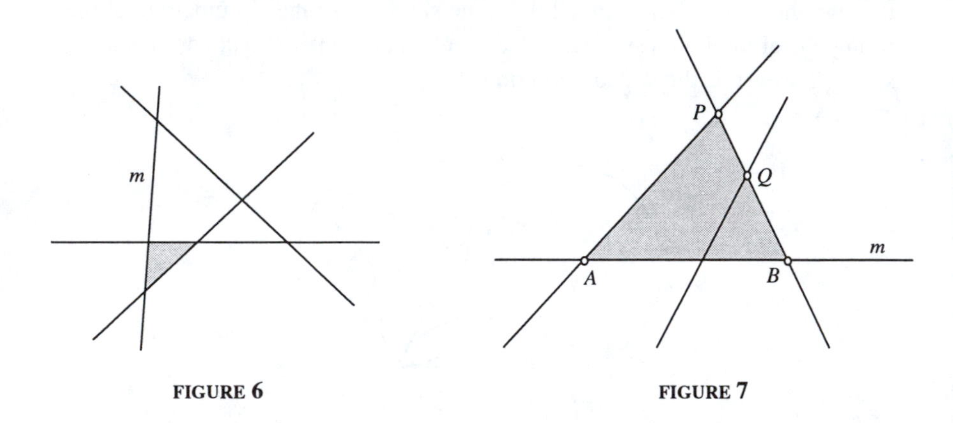

FIGURE 6 FIGURE 7

A problem like this gives one pause because there doesn't seem to be any way to come to grips with the number of sides of these regions. Again the key is to consider a minimum, in this case the point of intersection P of two lines in L which is closest to the line m under investigation (Figure 7). Since no three lines are concurrent, there are only two lines through P, and since no two lines are parallel, they each cross m to form a triangle PAB. Now, no other line of L could invade $\triangle PAB$ without creating a contradictory point of intersection Q which is closer to AB than P. Hence $\triangle PAB$ is one of the regions adjacent to m.

6. If a cube is cut up into a finite number of smaller cubes, prove that some two of them must be the same size.

Proceeding indirectly, suppose all the small cubes are different sizes. Now, these cubes can be packed in a box which is the same size as the original cube. Then the smallest cube C which rests on the bottom of the box will be completely overshadowed by the neighboring cubes. It is easy to see that C cannot be up against a side of the box without needing a cube of the same size, or smaller ones, to fill in the space around it (Figure 8).

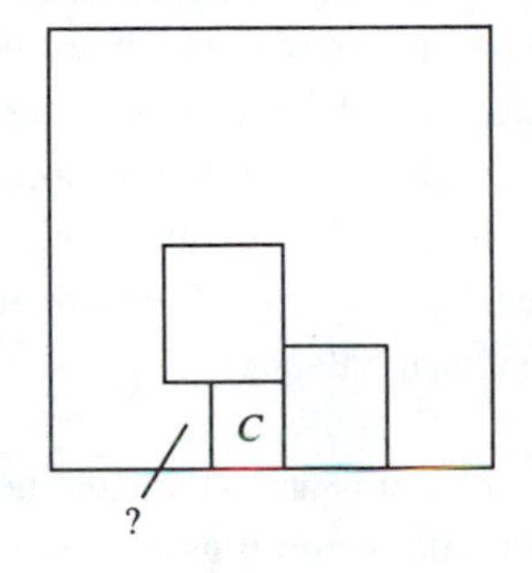

FIGURE 8

Thus C must occur somewhere in the interior of the floor of the box and its upper face F will be walled in on all sides (Figure 9). Since no two cubes are the same size, the cubes which touch F must all be smaller than C.

Similarly, the smallest cube which rests on F is completely walled in and the cubes on top of it must be smaller still. This argument can be continued ad

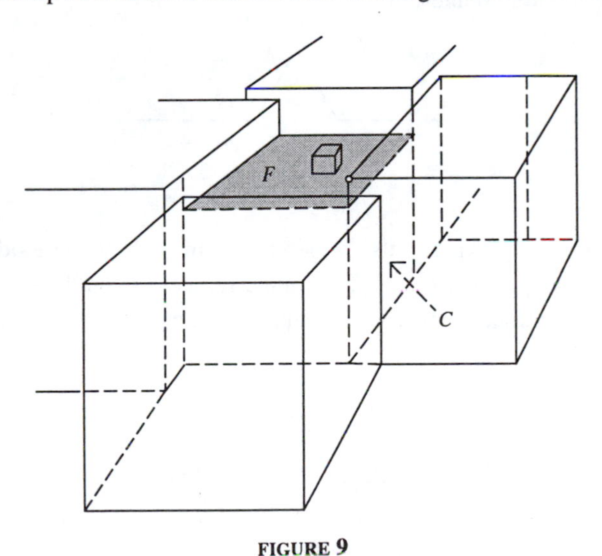

FIGURE 9

infinitum to yield the absurd conclusion that there is no smallest cube in the decomposition. Hence some two of the small cubes simply must be the same size.

7. Here is a brilliant twist on those problems about "3-weighings with an equal arm balance."

   At a trial, 14 coins were introduced as physical evidence. An expert had already examined the coins and determined that seven were genuine and seven counterfeit, and he knows which are which. While they all look the same to the untrained eye, the Judge is aware that all the genuine coins are the same weight and all the counterfeit coins also weigh the same as each other, but they are lighter than genuine coins.

   How can the expert justify his claims to the Judge with only three weighings with an equal arm balance?

The expert knows which coins are which and he must arrange to set them off against each other so that the weighings convey instructive messages to the Judge. Since it is only the Judge's reaction that counts, let us listen in on his thoughts as the weighings are presented to him.

First of all the expert puts a single coin on each side, X on the left and Y on the right, and the Judge observes that the right side goes down (Figure 10). Thus he says to himself that Y must be genuine and X counterfeit, and in order to keep track of things, he orders that Y be marked with a G for genuine and X be marked C for counterfeit.

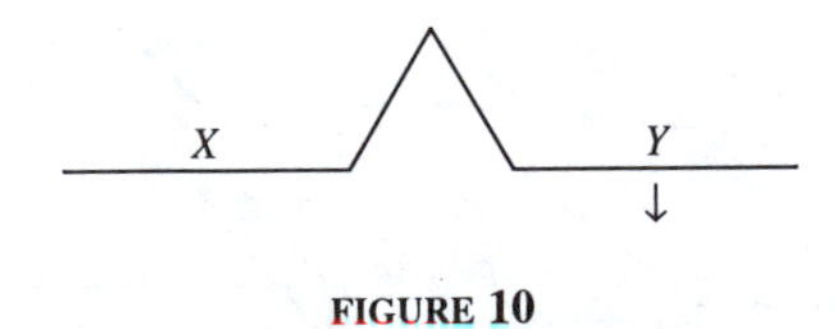

FIGURE 10

Next the crafty expert puts the genuine coin G on the left side with two unidentified coins P and Q, and the counterfeit coin C on the right side with two other unidentified coins S and T, and again the right side is observed to go down (Figure 11).

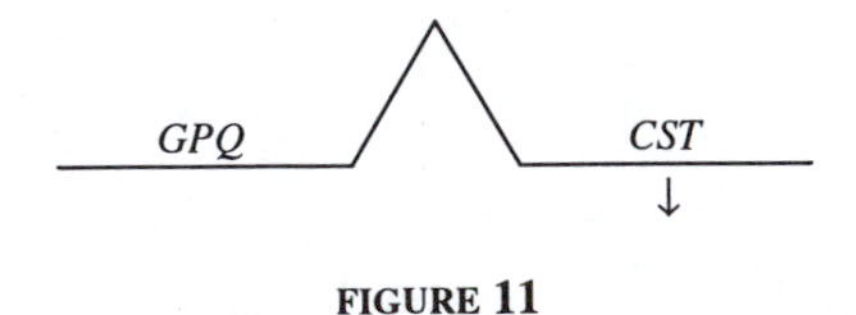

FIGURE 11

Now, it is clear that, when there is the same number of coins on each side, the side which goes down must contain more genuine coins than the other side. Thus the Judge acknowledges that the right side CST must contain more genuine coins than the left side GPQ. But C is counterfeit, and so the right side has at most two genuine coins in S and T. Thus the lighter left side can't have more than one genuine coin, and we conclude immediately that P and Q must both be counterfeit; and since the right side is in fact heavier, S and T must both be genuine. Thus the Judge marks S and T with Gs and P and Q with Cs, bringing the total number of identified coins to three of each kind.

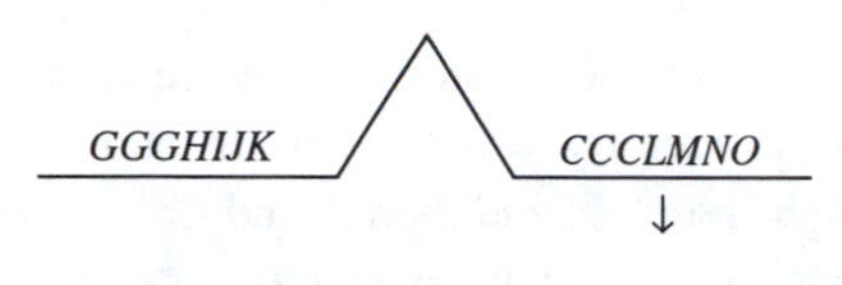

FIGURE 12

Finally, the expert places the three known genuine coins on the left with four unidentified coins to give $GGGHIJK$, and on the right he puts the three known counterfeit coins with the four remaining unidentified coins to give $CCCLMNO$ (Figure 12), and for the third time the Judge observes that the right side goes down. Thus he reckons that, since there are three known Gs on the left side, there must be at least four of them on the right side. Since L, M, N, O are the only possible genuine coins on the right, each of them must in fact be genuine. Not only that, but, if any of H, I, J, K were also genuine, there would be at least four genuine coins on the left side which the right side could not overbalance, and so each of H, I, J, K must be counterfeit, in complete justification of the expert.

8. Here is another engaging 3-weighings problem; it comes from the 1990 Tournament of the Towns for ages 13 to 15.

   In a set of 61 coins that look alike, 2 are counterfeit and the rest are genuine. The counterfeit coins weigh the same but their weight is different from the weight of a genuine coin. How can one determine in three weighings with an equal-arm balance whether a counterfeit coin is heavier or lighter than a genuine coin?

Even though this problem is accessible to elementary school pupils, its solution is delightfully ingenious.

First of all, throw away one of the coins; it may be counterfeit, but there is still at least one counterfeit in the remaining 60 coins. Now divide them into three groups A, B, C, of 20 coins each. Clearly, at least one of the groups must have a counterfeit coin, but not all three of them, and so the three groups cannot all be the same and there are essentially only two cases:

(i) two of A, B, C have all genuine coins and the third has one or both of the counterfeit coins,

(ii) two of A, B, C each have one counterfeit coin and the third has only genuine coins.

In any event, then, some two of the groups must have the same makeup: either both are entirely genuine or each has one counterfeit coin.

Now, it is easy to identify which of A, B, and C are the same in two weighings: first weigh A and B, and then A and C.

Clearly a balance immediately reveals the identical pair; and if neither weighing yields a balance, thus eliminating $A = B$ and $A = C$, the identical pair must be the only remaining possibility of B and C.

If necessary, let the groups be re-labelled so that A and B are the identical ones and C the individual group. Then the cases may be described as follows:

	A	B	C
Case 1	all genuine	all genuine	1 or 2 counterfeit
Case 2	1 counterfeit	1 counterfeit	all genuine

Now, the purpose of these weighings is not just to determine the identical groups A and B. If A and B are identified by the first weighing, we proceed with the second weighing anyway in order to find out whether C is heavier or lighter than A and B. Since the two weighings involve all three of the groups, when it takes two weighings to identify A and B, the weighing involving C directly reveals whether C is heavier or lighter than A and B. For definiteness, suppose that C is lighter. Then we can complete the description of the two cases:

	A	B	C
Case 1	all genuine	all genuine	1 or 2 counterfeit and C is lighter
Case 2	1 counterfeit	1 counterfeit	all genuine and C is lighter

For the third weighing we simply divide A in half and weigh the two halves against each other. Clearly, if they balance, case 1 must be valid, and if not, it must be case 2, for then one half must contain a counterfeit coin which the other doesn't. Since C is lighter than A, we immediately have the conclusions:

> in case 1: the counterfeit coin(s) in C is lighter than a genuine coin;
> in case 2: the counterfeit coin in A is heavier than a genuine coin.

9. $ABCD$ is a cyclic quadrilateral in which $AD = AB + CD$. Prove that the bisectors of angles B and C meet on AD (Figure 13).

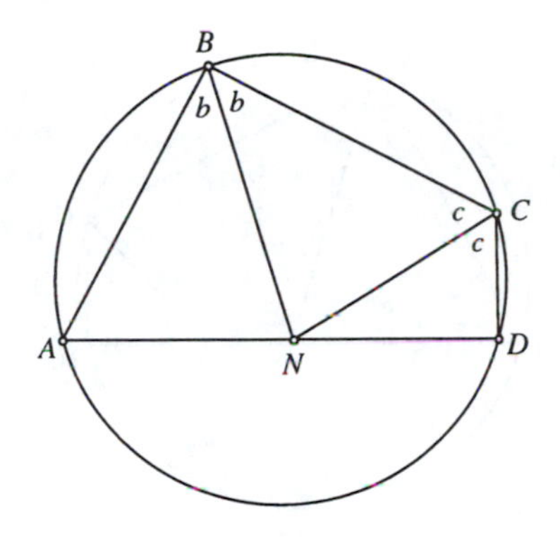

FIGURE 13

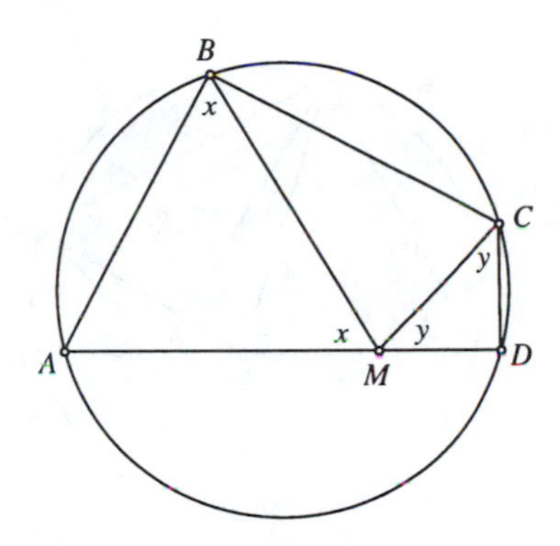

FIGURE 14

If M is marked on AD so that $AM = AB$, then since $AD = AB + CD$, it follows that $MD = CD$ and the triangles ABM and MCD are isosceles. Let the equal base angles at B and M be x and those at M and C be y (Figure 14).

Now,

$$\angle A + 2x = 180^\circ \quad \text{in } \triangle ABM,$$

and

$$\angle A + \angle C = 180^\circ \quad \text{in cyclic quadrilateral } ABCD.$$

Hence

$$\angle C = 2x,$$

and similarly,

$$\angle B = 2y.$$

Suppose $x > y$ and the bisector of $\angle C$ meets AD at N (Figure 15). Then

$$\angle BCN = \frac{1}{2}\angle C = x,$$

and the segment BN subtends angle x at each of M and C, implying $BNMC$ is a cyclic quadrilateral. In this case, in Figure 16,

exterior angle y at M = interior $\angle NBC$ at the opposite vertex B,

and since $\angle B = 2y$, then BN bisects $\angle B$ and the conclusion follows.

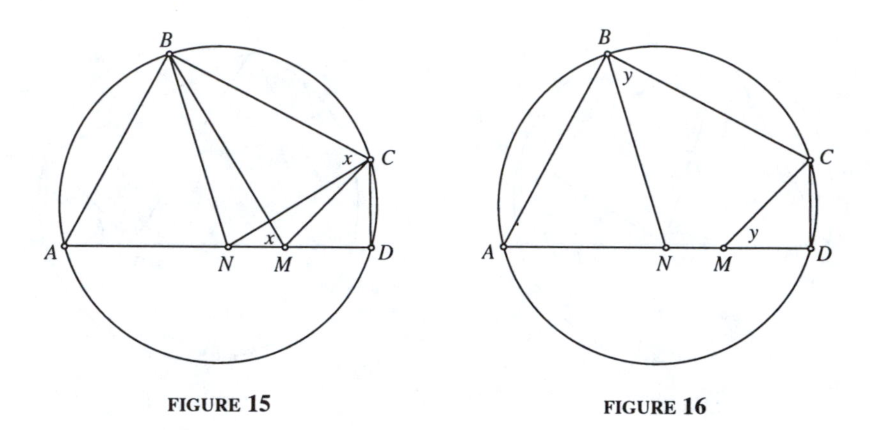

FIGURE 15 FIGURE 16

Two Properties of Quadrilaterals

10. Property One: If $ABCD$ is a quadrilateral that is circumscribed about a circle, the points of contact determine a second quadrilateral $PQRS$ (Figure 17). Prove that a pair of opposite sides of $PQRS$ and the diagonal of

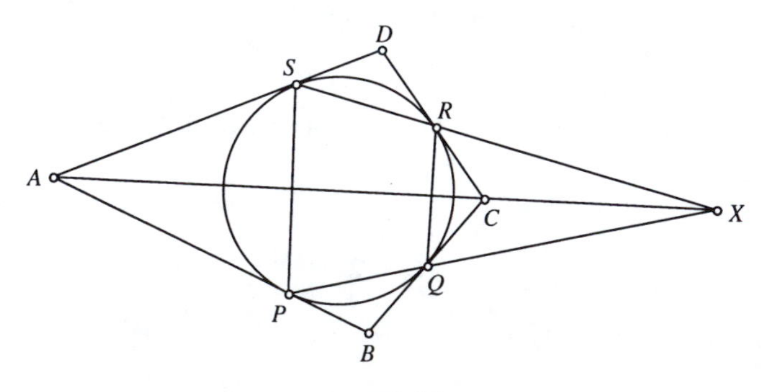

FIGURE 17

> $ABCD$ which lies between them are concurrent; for example, PQ, RS, and AC.

While one might expect to spend considerable effort to prove such a charming property, and be happy to do so, the result is an immediate consequence of the celebrated theorem of Menelaus.

If PQ and AC meet at X (Figure 18), we need to show that RS also goes through X. Since the two tangents to a circle from an external point have the same length, we have equal tangents as marked in Figure 18.

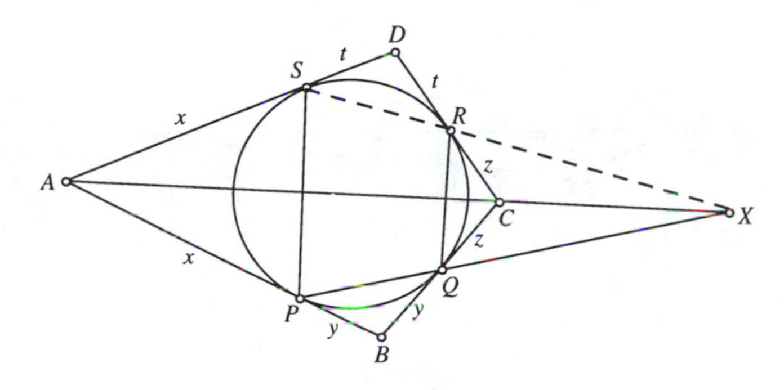

FIGURE 18

Now, PQX is a transversal across $\triangle ABC$, and so

$$\frac{y}{x}\cdot\frac{AX}{XC}\cdot\frac{z}{y}=-1,\quad\text{giving}\quad\frac{AX}{XC}=-\frac{x}{z}.$$

Conversely, S, R, and X are points on the sides of $\triangle ACD$ such that

$$\frac{t}{x}\cdot\frac{AX}{XC}\cdot\frac{z}{t}=\frac{AX}{XC}\cdot\frac{z}{x}=\left(-\frac{x}{z}\right)\cdot\frac{z}{x}=-1,$$

and it follows that S, R, and X are indeed collinear.

Now the case of parallel sides of $PQRS$ follows readily.

> If PQ and AC meet in a finite point X, then, as we have just seen, RS also goes through X, implying PQ and RS are not parallel. Thus, if PQ and RS were to be parallel, then PQ and AC could not meet in a finite point either, making all of PQ, AC, and RS parallel, that is, concurrent at infinity.

Therefore the property is valid in all cases.

Of course, this property holds for both pairs of opposite sides of $PQRS$. Now, if a quadrilateral $ABCD$ does not have an incircle, or if points P, Q,

R, S on its sides are not the points of contact of the incircle, then the property isn't generally valid. However, the following engaging property **does** hold for points P, Q, R, S, one on each side of any convex quadrilateral $ABCD$:

> Property Two: If one pair of opposite sides of $PQRS$ and the diagonal of $ABCD$ which lies between them are concurrent, then the same is true of the other pair (Figure 19).

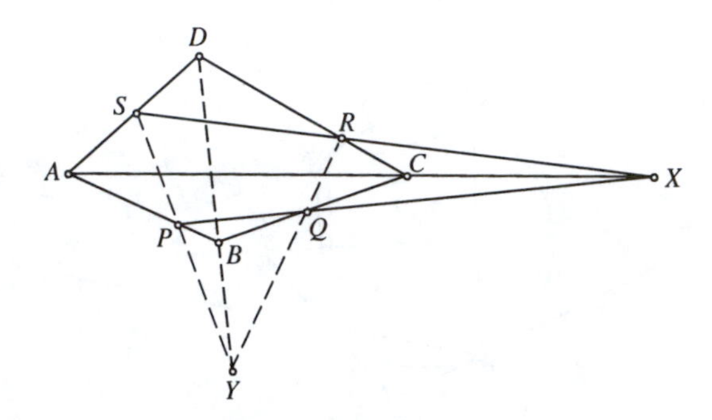

FIGURE 19

A detailed proof is given below, although it is hardly necessary to do more than point out that this is another simple application of the theorem of Menelaus.

Suppose PQ and RS meet at X on AC and that PS meets BD at Y (Figure 20); it remains to show that QR goes through Y. Let the lengths of the segments be labelled as in Figure 20. Then from the transversal SRX across $\triangle ACD$, we have

$$\frac{h}{i} \cdot \frac{AX}{XC} \cdot \frac{s}{t} = -1, \quad \text{giving} \quad \frac{AX}{XC} = -\frac{it}{hs};$$

similarly, from transversal PQX across $\triangle ABC$, we have

$$\frac{k}{j} \cdot \frac{AX}{XC} \cdot \frac{n}{m} = -1, \quad \text{giving} \quad \frac{AX}{XC} = -\frac{mj}{kn}.$$

Hence

$$\frac{it}{hs} = \frac{mj}{kn}, \quad \text{giving} \quad itkn = hsmj.$$

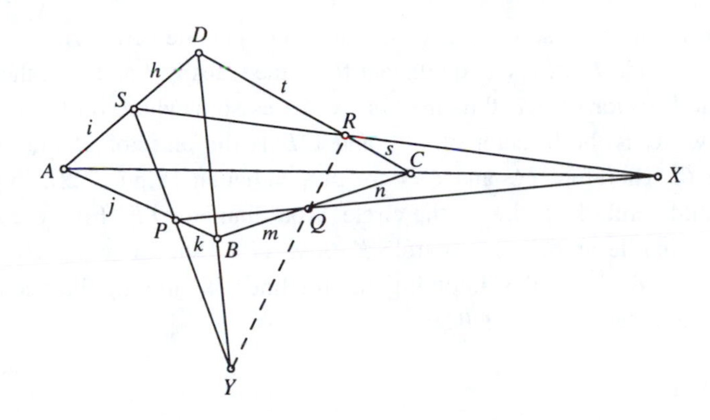

FIGURE 20

Again, from transversal SPY across $\triangle ABD$, we get

$$\frac{h}{i} \cdot \frac{j}{k} \cdot \frac{BY}{YD} = -1, \quad \text{implying} \quad \frac{BY}{YD} = -\frac{ik}{hj}.$$

It follows, then, that R, Q, and Y are points on the sides of $\triangle BCD$ such that

$$\frac{n}{m} \cdot \frac{BY}{YD} \cdot \frac{t}{s} = \frac{nt}{ms} \cdot \frac{BY}{YD} = \frac{nt}{ms} \cdot \left(-\frac{ik}{hj}\right)$$

$$= -\frac{itkn}{hsmj} = -1 \quad (\text{since } itkn = hsmj),$$

and the conclusion follows.

11. Construct $\triangle ABC$, given its circumcenter O, its incenter I, and an excenter E (Figure 21).

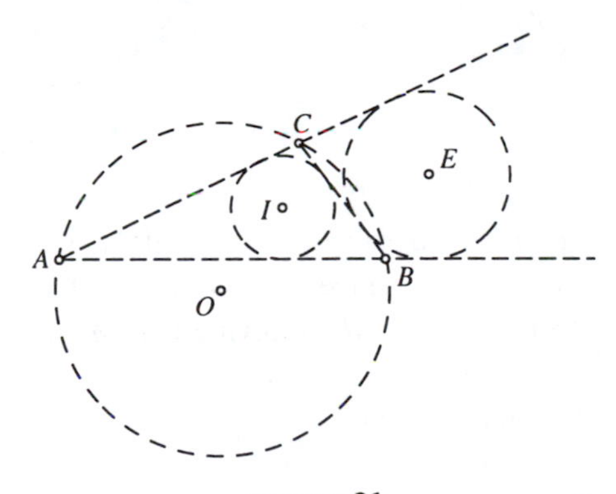

FIGURE 21

For definiteness, suppose E is the excenter opposite vertex A.

Since I and E are each equidistant from the sides AB and AC, they both lie on the bisector of $\angle A$; thus the vertex A lies somewhere on the line IE. Similarly, IC is the bisector of $\angle C$, and CE is the bisector of the exterior angle at C. Therefore IC and CE are perpendicular (Figure 22), implying that C, and similarly B, lie on the circle S on diameter IE (Figure 23). But B and C also lie on the circumcircle K of $\triangle ABC$, and therefore BC is the common chord of K and S, implying that the line OM joining their centers is the perpendicular bisector of BC.

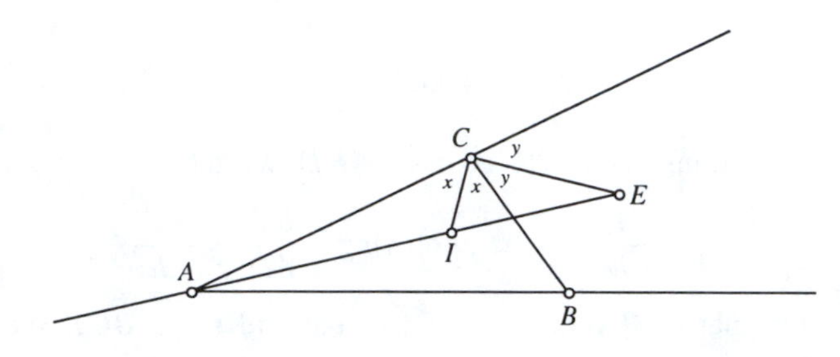

FIGURE 22

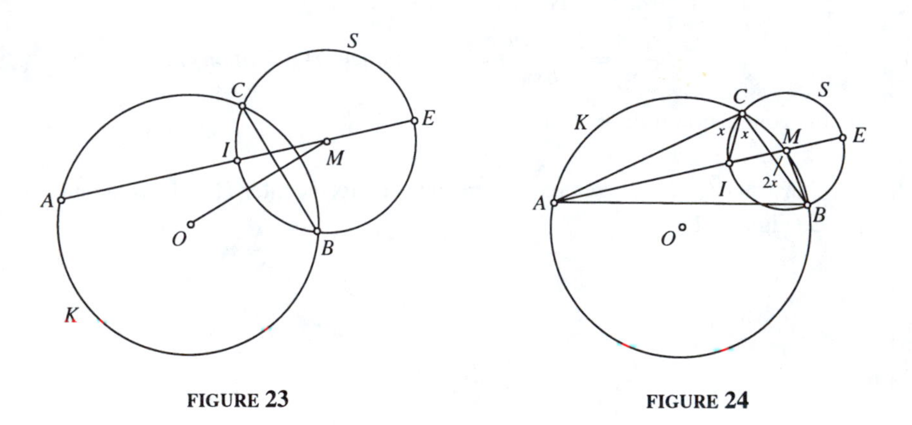

FIGURE 23 FIGURE 24

Now, if the chord IB of S subtends the angle x at C, it subtends $2x$ at the center M (Figure 24). But since IC bisects $\angle C$, then $\angle C = 2x$, and AB subtends $2x$ at each of C and M, implying that M must also lie on the circumcircle K.

Thus the construction is evident:

1. Bisect IE to get the midpoint M.

2. Construct the circle S on diameter IE and the circle K with center O and radius OM to get their points of intersection B and C.
3. Extend EI to meet K at A.

Proof: Clearly O is the circumcenter of $\triangle ABC$. Now, if it can be shown that I is the incenter, it would follow easily that E is the excenter opposite A:

> in that case, AI would be known to bisect $\angle A$, placing E on the bisector of $\angle A$, and since $\angle ICE$ is a right angle (in a semicircle of S), then E would bisect the exterior angle at C, making E the required excenter opposite A.

It remains, then, to show that I is the incenter.

Since the incenter of a triangle lies on the bisectors of all three angles, each side of the triangle subtends an angle at the incenter equal to $90° + \frac{1}{2}$(the opposite angle of the triangle) (Figure 25); thus another specification of the incenter is the point on the bisector of $\angle A$ at which BC subtends the angle $90° + \frac{A}{2}$:

In Figure 25,

$$2x + 2y + A = 180°, \quad \text{giving} \quad x + y = 90° - \frac{A}{2};$$

thus

$$\angle BIC = 180° - (x + y) = 90° + \frac{A}{2}.$$

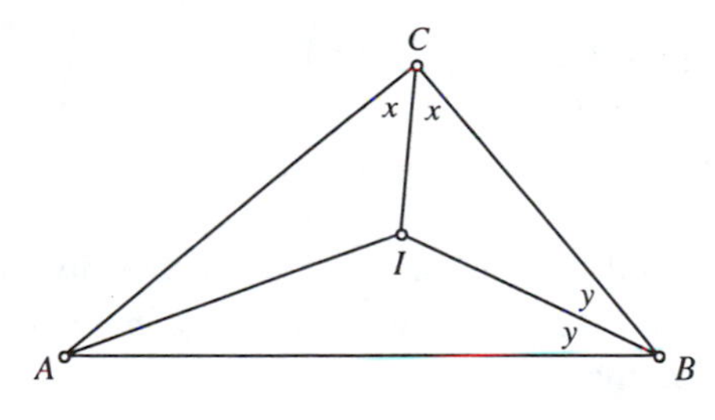

FIGURE 25

Now, in K, let

$$\angle CAM = \angle CBM = u,$$

and let

$$\angle BAM = \angle BCM = v.$$

But $CM = BM =$ the radius of S, implying $u = v$. Hence AI bisects $\angle A$.

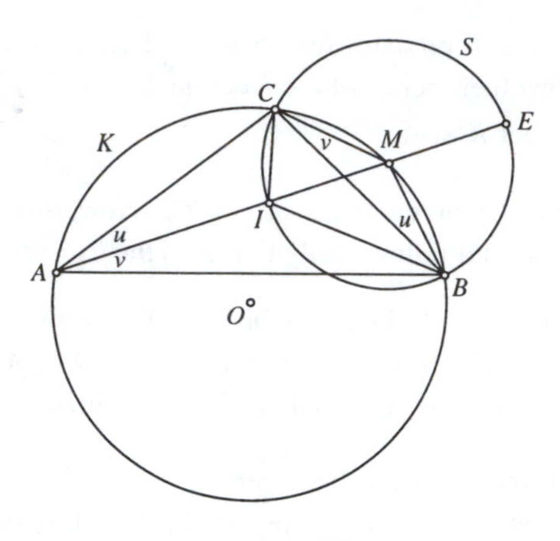

FIGURE 26

Finally, in cyclic quadrilateral $ABMC$,

$$\angle CMB = 180^\circ - \angle A,$$

implying that

$$\text{reflex } \angle CMB = 180^\circ + \angle A.$$

Since arc CEB of S subtends half the angle at I on the circumference as it does at the center M, then

$$\angle CIB = \frac{1}{2} \text{ reflex } \angle CMB = 90^\circ + \frac{A}{2},$$

completing the proof.

12. As usual, let $\sigma(n)$ denote the sum of the positive divisors of the positive integer n and $\varphi(n)$ the Euler φ-function. Prove that $\sigma(n)$ and $\varphi(n)$ always average at least n, that is,

$$\sigma(n) + \varphi(n) \geq 2n.$$

Let S be the set of integers left in $\{1, 2, \ldots, n\}$ when the integers counted by $\varphi(n)$ are taken away. For example, for $n = 15$, $\varphi(15)$ counts the 8 integers $\{1, 2, 4, 7, 8, 11, 13, 14\}$, making S the complementary set $\{3, 5, 6, 9, 10, 12, 15\}$. Because a member of S must share a common divisor $d > 1$ with n, it follows that S contains precisely those integers $\leq n$ which are themselves divisors $d > 1$ or multiples kd of such divisors. Since d is the first multiple of itself, the composition of S is described exactly by

$$S = \{\text{integers } kd \text{ which are multiples of a divisor } d \text{ of } n,\ d > 1\}.$$

Clearly, the **number** of integers left in S is simply $n - \varphi(n)$.

Now, the cardinality of S can also be determined by counting up these multiples kd. For each divisor d there are all the multiples $d, 2d, \ldots$ which do not exceed n; since d divides n, there are $\frac{n}{d}$ of them. If the divisors of n are

$$d_1 = 1 < d_2 < d_3 < \cdots < d_{k-1} < d_k = n,$$

then, discarding the divisor 1 since S contains only integers > 1, S would contain the $\frac{n}{d_2}$ multiples of d_2, the $\frac{n}{d_3}$ multiples of d_3, and so forth, for a total of

$$T = \frac{n}{d_2} + \frac{n}{d_3} + \cdots + \frac{n}{d_k}.$$

However, the same integer can be a multiple of more than one of the divisors (for $n = 18$, 6 is a multiple of the three divisors 2, 3, and 6), and so in this sum the same integer might be counted more than once. Therefore, if not exactly right, T is an overestimate of the cardinality of S and we have

$$|S| \le T.$$

It remains only to calculate T.

We begin by observing that if d is a divisor of n, so is $\frac{n}{d}$; in fact the divisors $\frac{n}{d_2}, \frac{n}{d_3}, \ldots, \frac{n}{d_{k-1}}$ are merely the divisors $d_2, d_3, \ldots, d_{k-1}$ in reverse order. Thus, not forgetting to include the final term $\frac{n}{d_k}$ $(= \frac{n}{n} = 1)$ in the expression for T, we have

$$\begin{aligned} T &= d_2 + d_3 + \cdots + d_{k-1} + 1 \\ &= [(d_1 + d_2 + \cdots + d_k) - d_1 - d_k] + 1 \\ &= [\sigma(n) - 1 - n] + 1 \\ &= \sigma(n) - n. \end{aligned}$$

Therefore

$$|S| \le T$$

yields

$$n - \varphi(n) \le \sigma(n) - n,$$

and the desired

$$2n \le \sigma(n) + \varphi(n).$$

13. Prove that, in any covering of a 99×99 checkerboard with tiles of the following three kinds, at least 199 of the L-shaped tiles must be used.

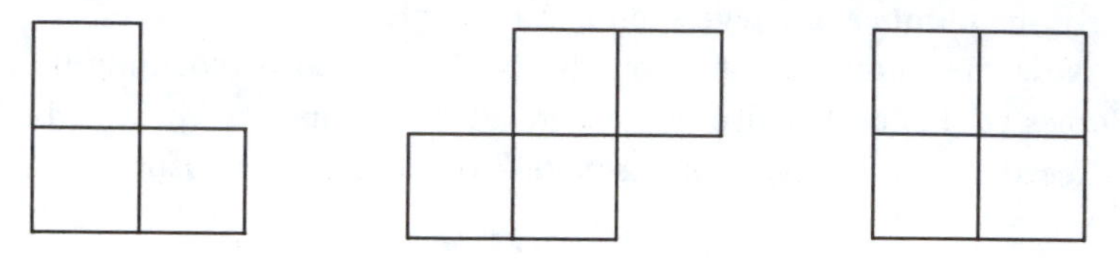

FIGURE 27

Let the odd-numbered squares in the odd-numbered columns be colored red (Figure 28). Thus 50 squares in each of 50 columns are colored for a total of $50^2 = 2500$.

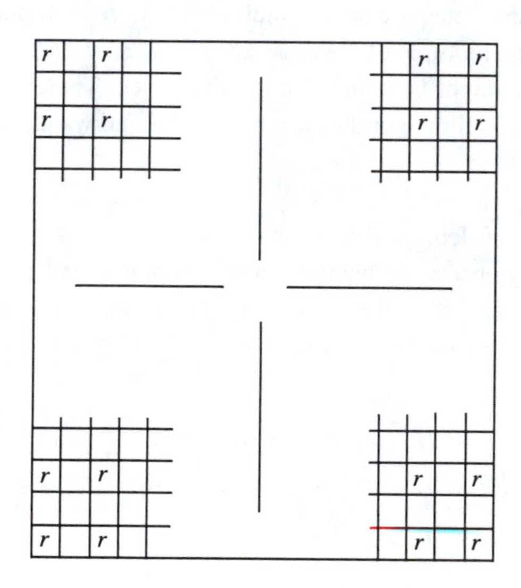

FIGURE 28

Clearly no tile of any kind is capable of covering more than one red square, and it is possible for an L-shaped tile to miss them completely. Thus a total of at least 2500 different tiles are needed to cover the board. If there are m L-shaped tiles and altogether n tiles of the second and third kinds (i.e., the 4-cell tiles), then

$$m + n \geq 2500. \tag{1}$$

Now, each L-shaped tile covers 3 squares of the board and each other tile 4 squares, and so the total number of squares covered is

$$3m + 4n = 99^2 = 9801. \qquad (2)$$

From 2, we get

$$4n = 9801 - 3m,$$

and from 1, that

$$4m + 4n \geq 10000.$$

Hence

$$4m + (9801 - 3m) \geq 10000,$$

and

$$m \geq 199, \qquad \text{as desired.}$$

This argument generalizes directly to any odd-sized board: a $(2k-1) \times (2k-1)$ board requires at least $4k - 1$ L-shaped tiles.

14. The integers $\{1, 2, 3, \ldots, 2n\}$ are divided arbitrarily into two subsets A and B of n integers each. The numbers in A are arranged in increasing order and those in B in decreasing order:

$$A = \{a_1 < a_2 < \cdots < a_n\},$$
$$B = \{b_1 > b_2 > \cdots > b_n\}.$$

Prove that the sum of the corresponding positive differences

$$S = |a_1 - b_1| + |a_2 - b_2| + \cdots + |a_n - b_n|$$

is always equal to n^2.

For example, for $n = 4$, we might have

$$A = \{1, 3, 6, 7\},$$
$$B = \{8, 5, 4, 2\},$$

giving the sum

$$7 + 2 + 2 + 5 = 16 = 4^2.$$

The solution hinges on the brilliant observation that, in each positive difference $|a_k - b_k|$, the two integers straddle $n + \frac{1}{2}$; that is to say, the greater of them is always from the subset $\{n+1, n+2, \ldots, 2n\}$ and the lesser from $\{1, 2, \ldots, n\}$. This is easily proved indirectly:

> Suppose, for definiteness, that both a_k and b_k are $\leq n$. In A, there are $k-1$ integers that are smaller than a_k, namely $a_1, a_2, \ldots, a_{k-1}$, and in B, the $n-k$ integers $b_{k+1}, b_{k+2}, \ldots, b_n$ are smaller than b_k, for a total of $(k-1)+(n-k)+2 = n+1$ (counting a_k and b_k themselves) positive integers $\leq n$, which is impossible.

Clearly $|a_k - b_k|$ = the greater integer − the lesser, and over the n differences the greater integer runs through all the values $n+1, n+2, \ldots, 2n$, while the lesser integer takes all the values $1, 2, \ldots, n$. Thus, gathering all the positive greater integers together and all the negative lesser integers together, we have

$$\begin{aligned} S &= [(n+1)+(n+2)+\cdots+2n] - [1+2+\cdots+n] \\ &= [(n+1)-1]+[(n+2)-2]+\cdots+[(n+n)-n] \\ &= n+n+\cdots+n \text{ (n times)} \\ &= n^2. \end{aligned}$$

15. D and E are points of trisection on the sides AC and AB of equilateral triangle ABC, and BD and CE meet at P (Figure 29). Prove $\angle APC$ is a right angle.

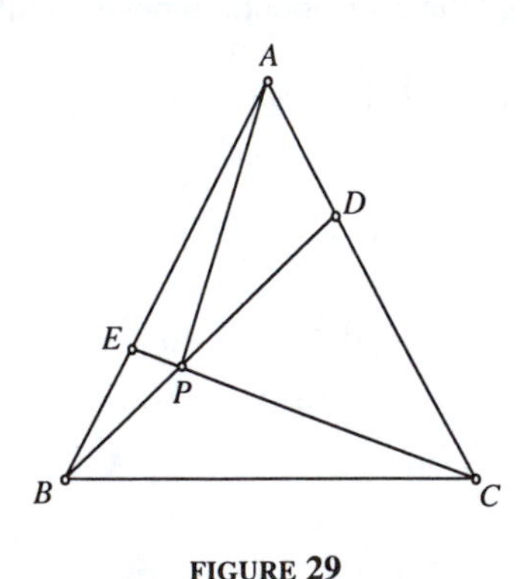

FIGURE 29

Since D and E are points of trisection, the triangles AEC and BDC are congruent (SAS):

$$AC = BC, \quad AE = \frac{2}{3} \text{ (of a side of } \triangle ABC) = DC, \quad \angle A = \angle C = 60^\circ.$$

Thus $\angle AEC = \angle BDC$, that is, the exterior angle of quadrilateral $AEPD$ at D is equal to the interior angle at E, and it follows that the quadrilateral is cyclic.

Now, in $\triangle AED$, EA is twice side AD and the angle between them is 60°, making it a 30°-60°-90° triangle with the right angle at D. Thus the circle on diameter EA goes through D, and since $AEPD$ is cyclic, it also goes through P. The diameter EA, then, also subtends a right angle at P and the conclusion follows.

16. Two Puzzle Games

(a) A game is played on a 4×4 board on which plus and minus signs are arranged as in Figure 30(a). There are three kinds of move in the game: reversing all the signs (i) in a row, (ii) in a column, (iii) along a diagonal or one of the shorter lines which are parallel to a diagonal.

Thus the first move might be of the third kind and result in Figure 30(b).

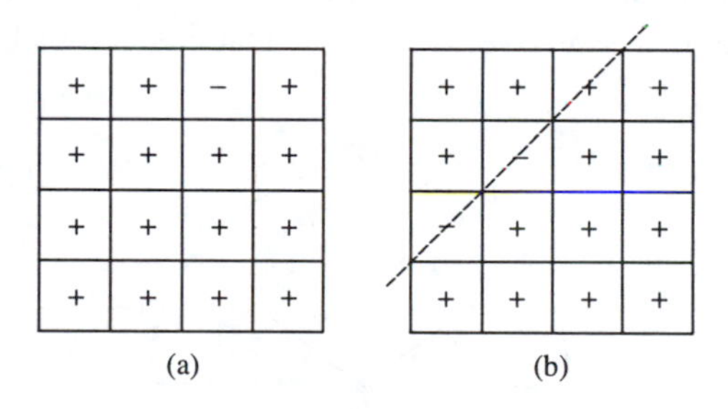

FIGURE 30

Either determine a sequence of moves to clear the board of minus signs or prove that it cannot be done.

Suppose each plus sign is replaced by $+1$ and each minus sign by -1, and consider the 8 squares in the shaded region in Figure 31. Clearly a move either leaves the shaded squares completely untouched or it reverses the signs of the numbers in exactly two of them. Therefore the **product** of the numbers in the shaded region is always the same, namely the $(+1)^7(-1) = -1$ with which it begins the game. Thus an odd number of shaded squares must always hold -1, i.e., at least one square, implying it is impossible to get rid of all the minus signs on the board.

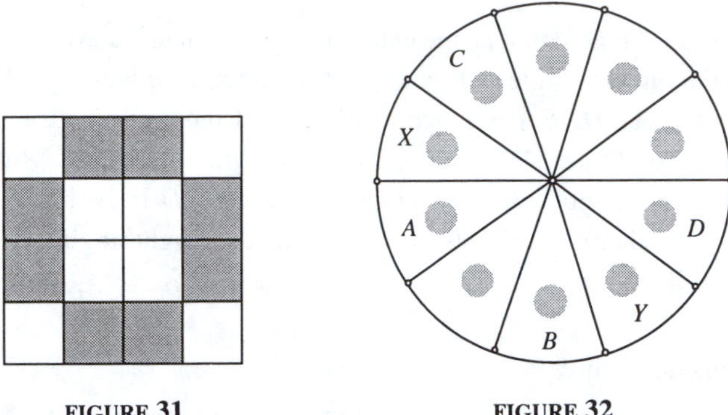

FIGURE 31 FIGURE 32

(b) A game is played on a circle which is divided into ten sectors in each of which a chip is placed (Figure 32). A move consists of selecting two chips and moving each to an adjacent sector; for example, for the pair (X, Y), X could be moved to A or to C and Y to B or D.

Either determine a sequence of moves which brings all the chips together in the same sector or prove that it is impossible to do so.

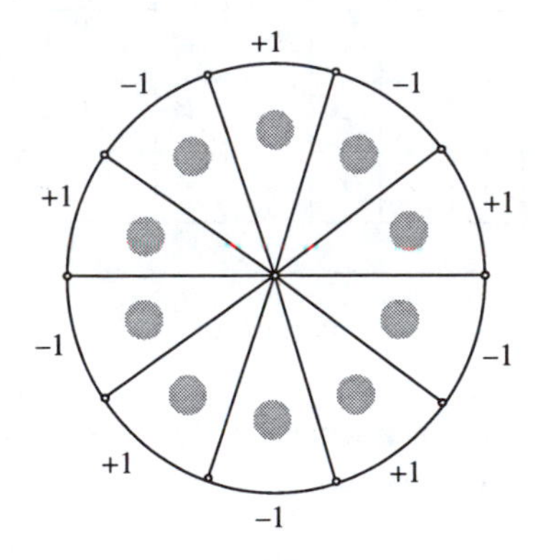

FIGURE 33

Let the sectors be marked alternately $+1$ and -1 (Figure 33), and throughout the game let each chip be given the value of the sector in which it resides, changing its value as the chip is moved about. Thus a move reverses the signs of the values of two of the chips, keeping the **product** of the values of the ten chips constantly at the $(+1)^5(-1)^5 = -1$ at which it begins the game. Hence it is impossible to gather all the chips into one sector, for the product at such a position would be $+1(= (+1)^{10}$ or $(-1)^{10})$.

17. The circumference of a circle C is partitioned into $3k$ arcs: k of length 1, k of length 2, and k of length 3. Prove that, no matter what their order around C, some two of their endpoints will be diametrically opposite.

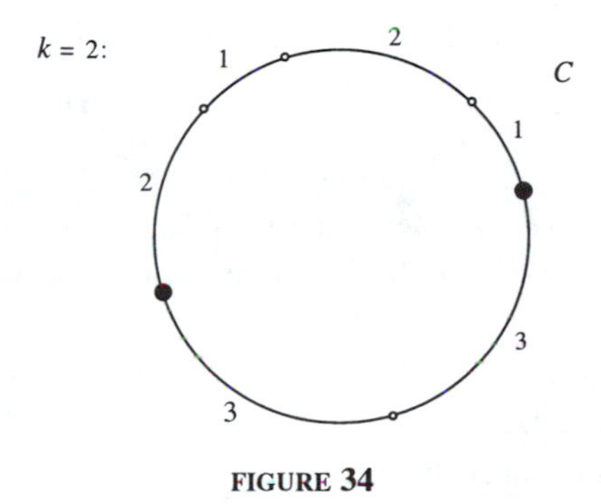

FIGURE 34

This intriguing problem was posed by V. Proizvolov and the following beautiful solution is due to the ingenious Vladimir Dubrovsky of Moscow State University.

Let the $3k$ endpoints of the arcs be colored white and let's color black the midpoints of the arcs of length 2 and the points of trisection of the arcs of length 3. Thus the white and black points divide the circumference into $6k$ unit arcs and accordingly are the vertices of a regular $6k$-gon inscribed in C. Since $6k$ is even, these vertices go together into diametrically opposite pairs, and since $3k$ of the points are white, the other $3k$ must be black. We want to prove that some pair of diametrically opposite vertices are both white. Proceeding indirectly, suppose that each of the $3k$ pairs of opposite vertices consists of a white point and a black one. Professor Dubrovsky now argues brilliantly to a contradiction as follows.

The endpoints of an original arc AB of length 1 are both white and hence they must be opposite a pair of consecutive black points C and D, which can occur only as the two points of trisection of an original arc PQ of length 3 (Figure 35).

Now let each of the unit arcs AB and CD be contracted to a point, a double point, reducing the total length of the arcs to $6k - 2$ units, and let the

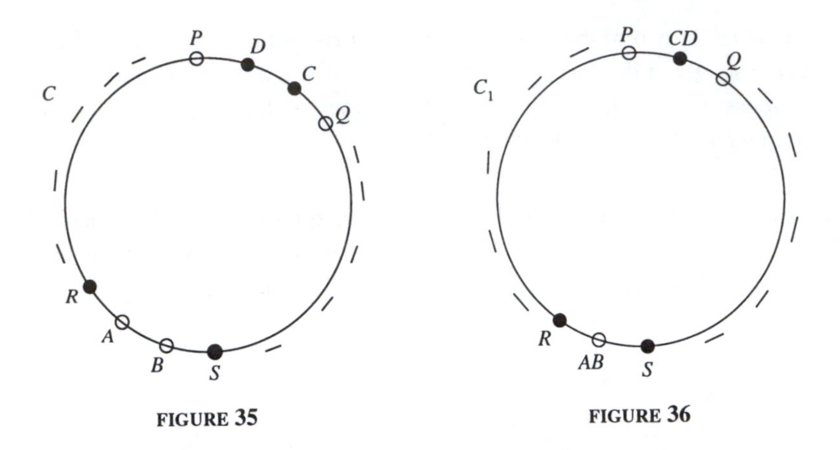

FIGURE 35 FIGURE 36

resulting arcs, which we consider to be flexible, be transferred, in order, to cover the circumference of a circle C_1 whose circumference is $6k - 2$:

> From the double vertex CD, representing both C and D, lay off in one direction the arcs $DP \ldots RA$ of C (Figure 36); clearly this carries one halfway around C_1 since the total lengths of these arcs is $3k-1$. Similarly, laying off in the other direction from CD the arcs $CQ \ldots SB$ of C would cover the other half of C_1 and bring one around to a vertex AB which represents both A and B. Let the point CD continue to be colored black and the point AB white.

Thus on C_1, the black vertex CD is diametrically opposite the white one AB and, in fact, each of the other points is opposite the same point it was opposite on C: P and S are still opposite, as are Q and R, ... That is to say, this operation of going from C to C_1 **preserves the unlike colors of the diametrically opposite pairs**, and so on C_1 every point is still opposite one of the other color.

In this process, we have lost entirely an arc AB of length 1 and converted an arc PQ of length 3 into one of length 2. Around C_1, then, there are $k - 1$ arcs of length 1, $k + 1$ arcs of length 2, and $k - 1$ arcs of length 3. Clearly, this operation can be continued to yield a sequence of k decreasing circles $C_1, C_2, \ldots, C_k$, at which point there are no arcs left of lengths 1 or 3 and the number of arcs of length 2 has grown to $2k$. That is to say, around the final circle C_k, there are $2k$ white points spaced equally at arcs of length 2; of course, the same is true of their $2k$ black midpoints, but this is irrelevant. Thus these $2k$ white points are the vertices of a regular inscribed $2k$-gon in C_k, and since $2k$ is even, these white points go together into k diametrically opposite pairs. Thus we have the contradiction that the unlike coloring of opposite pairs is **not** preserved on C_k, and the argument is complete.

18. $ABCDE$ is any zig-zag path across a circle such that the angles at B, C, and D are each 45° (Figure 37). Prove that the shaded and unshaded regions have the same area.

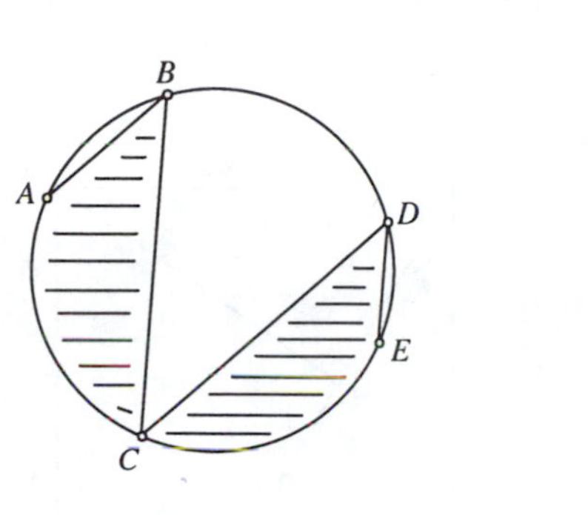

FIGURE 37

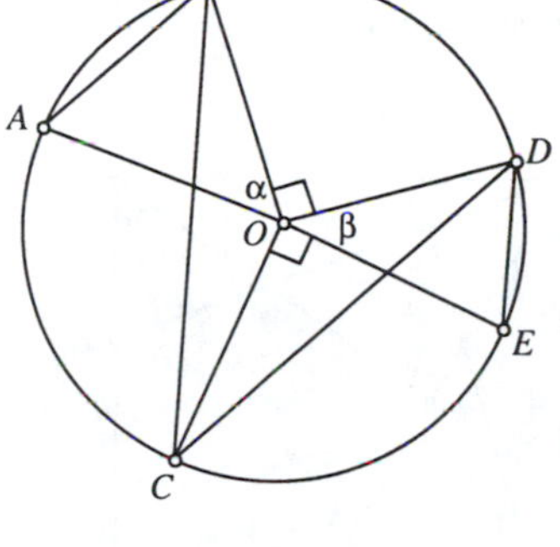

FIGURE 38

Since arcs AC, BD, and CE subtend 45° angles at the circumference, each must be one-quarter of the circumference, implying that AE is a diameter (Figure 38). Also, the quarter-circumference BD can't occur in the interior of semicircular arc $ABDE$ without having BC and CD straddle the center O. Thus the radii to A, B, C, D, E partition the circle as shown in Figure 38.

Let the radius of the circle be r, $\angle AOB = \alpha$ and $\angle DOE = \beta$. Now, the arcs BD and CE subtend right angles at the center, and so

$$\alpha + \beta + 90° = 180°,$$

implying angles α and $(\beta+90°)$ are supplementary; in particular, angles AOB and COD are supplementary. Thus triangles AOB and COD have the same area:

$$\Delta AOB = \frac{1}{2}r^2 \sin \angle AOB = \frac{1}{2}r^2 \sin \angle COD = \Delta COD.$$

Similarly, angles $(\alpha+90°)$ and β are supplementary, in particular angles BOC and DOE, leading to $\Delta BOC = \Delta DOE$.

Therefore the total unshaded area is given by

$$\begin{aligned}&(\text{the segments on } AB \text{ and } DE) + \text{the quarter circle } BOD\\ &+ \Delta BOC + \Delta COD\end{aligned}$$

$$= (\text{the segments on } AB \text{ and } DE) + \text{the quarter circle } BOD$$
$$+ \triangle DOE + \triangle AOB,$$

which clearly adds up to the semicircle ABE, and the conclusion follows.

19. An arbitrary pair of straight lines AX and AY are drawn across square $ABCD$, and perpendiculars are dropped from B and D to meet them at K, L, M, N (Figure 39). Prove that KL and MN are always equal and perpendicular.

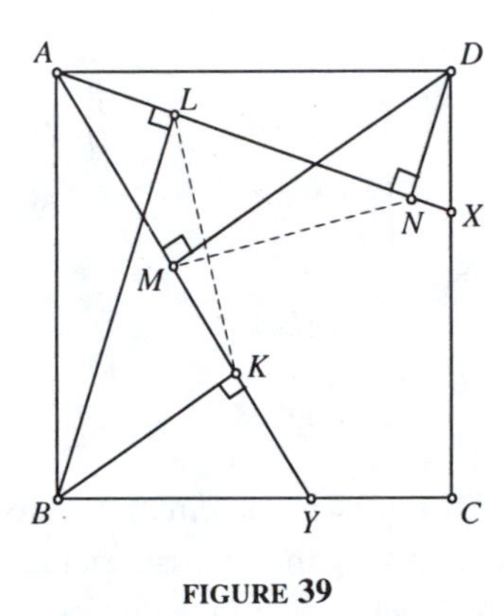

FIGURE 39

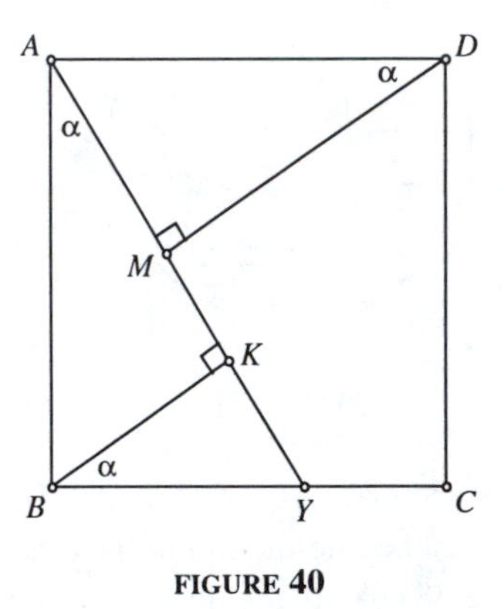

FIGURE 40

With the considerable symmetry of a square, it is always a good idea to check whether an appropriate quarter-turn, half-turn, or reflection might be of help. Accordingly, let the square be rotated clockwise about its center through an angle of 90°. This carries BC to AB and it is not difficult to see that K is carried to M (Figure 40): since

$$\angle KBY + \angle KBA = \text{the right angle at } B,$$

and also

$$\angle BAK + \angle KBA = 90°(\text{in right triangle } ABK),$$

it follows that

$$\angle KBY = BAK (= \alpha).$$

Similarly $\angle BAK = \alpha = \angle ADM$. Thus the rotation carries BK to run along AY, and AK to run along DM, and so the image of K must be the point of intersection M of AY and DM. Similarly the rotation carries L to N. Thus KL is carried to MN and the conclusion follows.

26 Intriguing Exercises

Virtually all of these problems have been taken from the two regular columns of *Quantum*—"Brainteasers," and "Challenges in Physics and Math," and are stated here without solution. In a few cases, however, a hint is offered at the very end of the essay. Specific references for these exercises are given immediately following the last exercise.

1. Boat 1 and boat 2, which travel at constant speeds, not necessarily the same, depart at the same time from docks A and B, respectively, on the banks of a circular lake. If they go straight to docks C and D, respectively, they collide. Prove that if boat 1 goes instead to dock D and boat 2 to dock C, they arrive simultaneously.
2. In Figure 41, $ABCD$ is a square. Prove that the sum of the areas $a + b + c = d$.

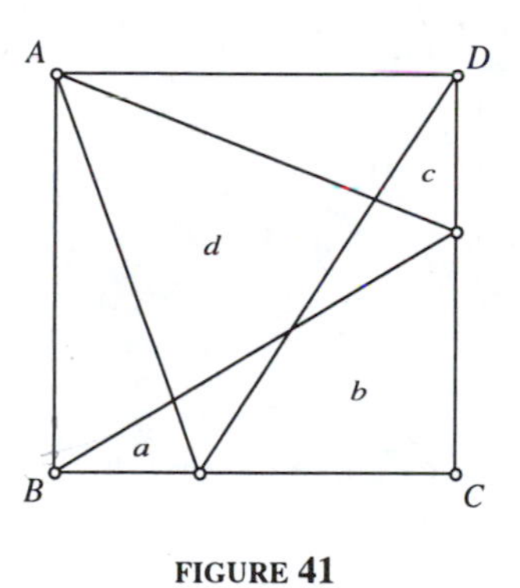

FIGURE 41

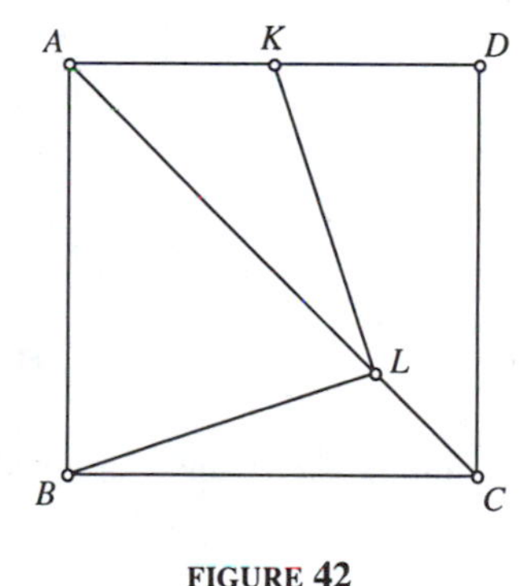

FIGURE 42

3. The point L divides the diagonal AC of square $ABCD$ in the ratio 3:1, and K is the midpoint of AD (Figure 42). Prove that $\angle BLK$ is a right angle.
4. In Figure 43, prove that the shaded region is half the area of the star.
5. In a circle C, two circles A and B are internally tangent to it at P and Q (Figure 44). If PQ goes through one of the points of intersection X of A and B, prove that the radii of A and B add up to the radius of C.

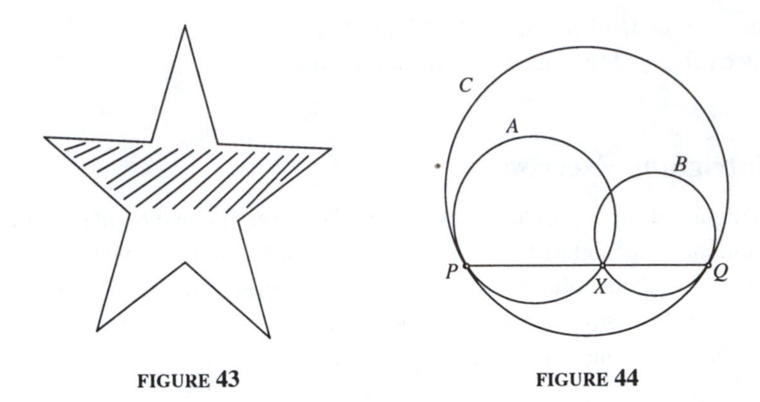

FIGURE 43

FIGURE 44

6. In Figure 45, squares are drawn on the sides of a right angled triangle. Prove that the triangles a, b, and c all have the same area.

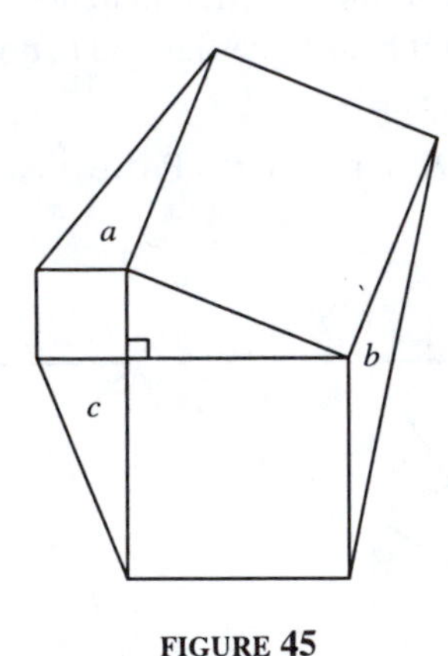

FIGURE 45

7. The numbers a, b, and c are positive real numbers less than $\frac{\pi}{2}$. If

$$\cos a = a,$$
$$\sin(\cos b) = b,$$

and

$$\cos(\sin c) = c,$$

arrange a, b, and c in order of magnitude.

8. Prove that, if one copy of a convex quadrilateral is cut along one diagonal and a second copy is cut along the other diagonal, the four triangles thus obtained can always be arranged to form a parallelogram.

9. If each point on a circle is colored either red or blue, prove that some three points of the same color determine an **isosceles** triangle.
10. Arrange the integers from 1 to 15 in a row so that the sum of every pair of adjacent integers is a perfect square.
11. Find the smallest positive integer n such that the sum of the digits of n and of $n+1$ are both divisible by 17.
12. For any positive real numbers a and b, prove that

$$2\sqrt{a}+3\sqrt[3]{b}=5\sqrt[5]{ab}.$$

13. Solve the equation $x^3+x^2+x=-\frac{1}{3}$.
14. If the sum of two positive real numbers is less than their product, prove that their sum must exceed 4.
15. In Figure 46, $AC = BD$ and angles ABC and BCD are supplementary. Prove the angles at A and D are equal.

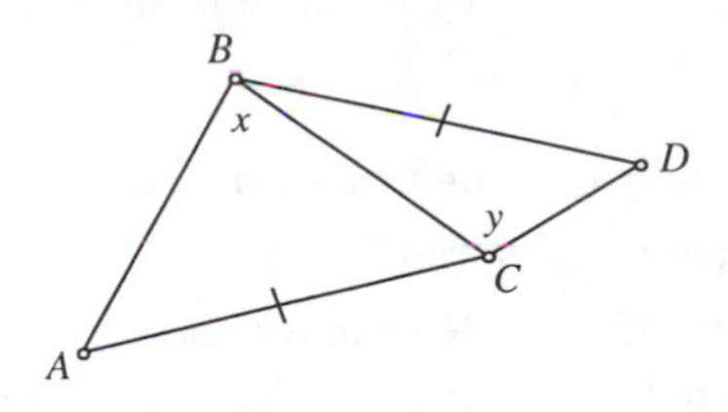

FIGURE 46

16. In a set of 1993 coins, 20 are known to be counterfeit. Now, one kind of counterfeit coin is known to be 1 gm. lighter than a genuine coin and an other kind 1 gm. heavier. How many there are of each kind is unknown. You have a pan balance with a pointer that registers the number of grams in the difference between the weights in its pans. Determine how to tell, in one weighing with such a balance, whether a single coin C selected from the set is genuine or counterfeit.
17. At a party each boy danced with three girls and each girl danced with three boys. Prove that the number of boys at the party was equal to the number of girls.
18. Two squares are inscribed in a semicircle as in Figure 47. Prove that the area of the big square is four times that of the little square.
19. The sum of two positive integers is 30030. Prove that their product is not divisible by 30030.

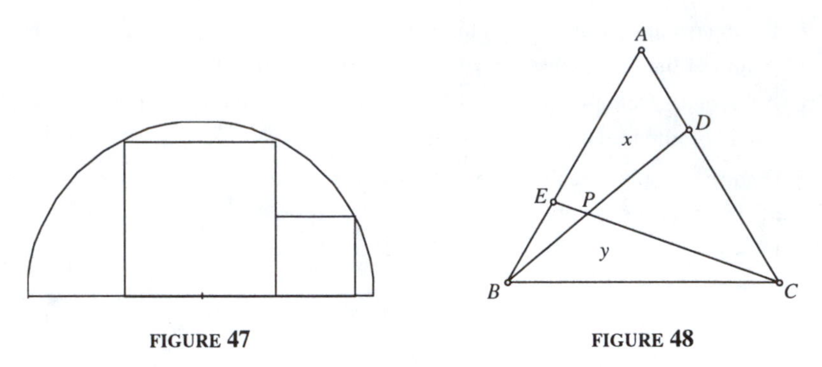

FIGURE 47

FIGURE 48

20. Lines BD and CE across equilateral triangle ABC meet at P as shown in Figure 48. If the areas of regions x and y are the same, determine $\angle BPC$.

21. In a strange notebook, the following one hundred statements were written:

    1. There is exactly one wrong statement in this book.
    2. There are exactly two wrong statements in this book.
    3. There are exactly three wrong statements in this book.

    $\vdots$

    100. There are exactly one hundred wrong statements in this book.

    Which of these statements is true?

22. The figure in Figure 49 is made from a square and a circular arc, of radius equal to the diagonal of the square, which spans a chord of length equal to twice a side of the square. Determine how to cut the figure into two congruent pieces.

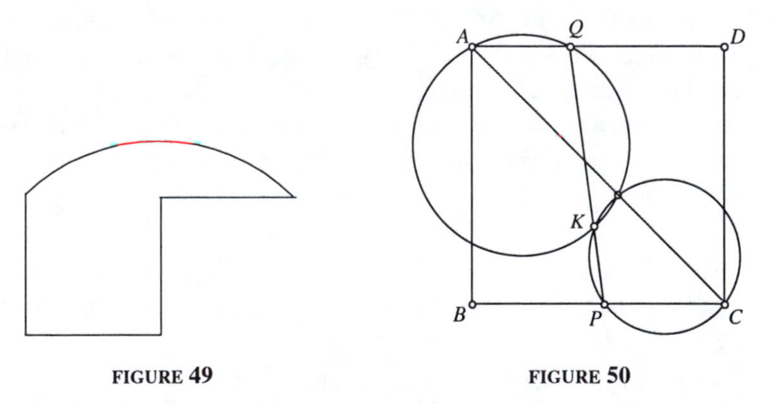

FIGURE 49

FIGURE 50

23. K is any point inside square $ABCD$ and PQ is any transversal through K (Figure 50). Prove that the circles around triangles AQK and PCK intersect on the diagonal AC.

24. There are three ordinary electric lamps in one room and three switches in another room. To begin, each lamp is off and each switch in the "off" position. Each switch is connected to exactly one of the lamps. How can you determine which switch is connected to which lamp if you are allowed to go into the room with the lamps only once? (No peeking through doorways, etc.)
25. A small square is place arbitrarily inside a big square (Figure 51). Prove that the areas

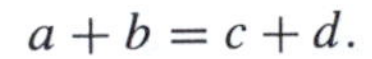

$$a + b = c + d.$$

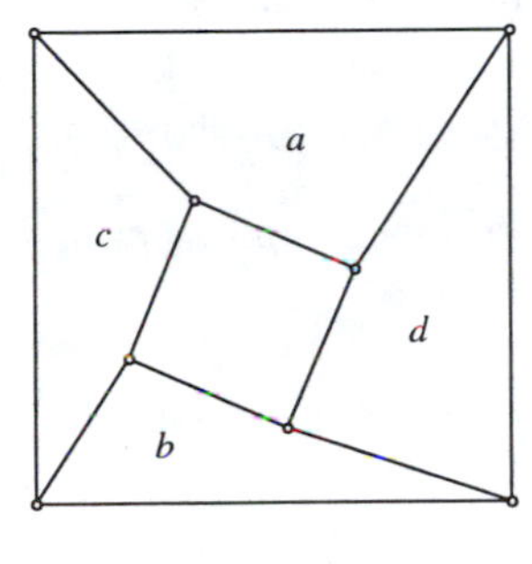

FIGURE 51

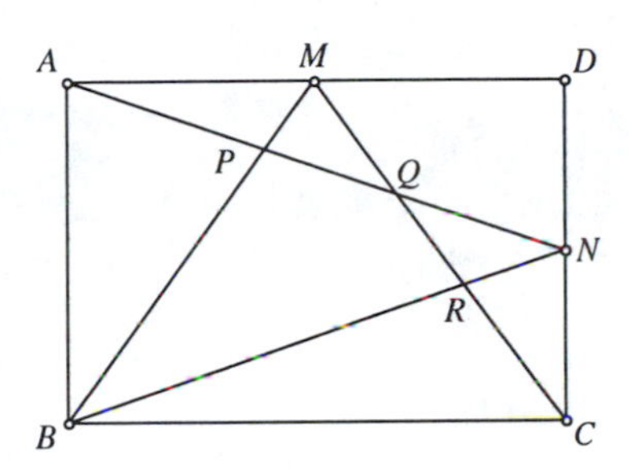

FIGURE 52

26. M and N are the midpoints of sides AD and CD in rectangle $ABCD$ (Figure 52). Prove that $PQRB$ is a cyclic quadrilateral.

References for the Exercises

1. Problem M26, March-April, 1991, 30; contributed by N. Vasilyev.
2. Problem B32, September-October, 1991, 21; contributed by V. Proizvolov.
3. Problem M36, November-December, 1991, 18; contributed by Y. Bogaturov.
4. Problem B48, March-April, 1992, 20; contributed by N. Avilov.
5. Problem M53, May-June, 1992, 19; contributed by A. Vesyolov.
6. Problem B54, May-June, 1992, 29; contributed by N. Avilov.
7. Problem M64, September-October, 1992, 16; contributed by S. Gessen.
8. Problem B62, September-October, 1992, 31; contributed by V. Proizvolov.
9. Problem B64, September-October, 1992, 31; contributed by I. Tonov (Bulgaria).
10. Problem B65, September-October, 1992, 31; contributed by B. Recamán (Swaziland).
11. Problem B69, November-December, 1992, 18; contributed by G. Galperin.
12. Problem M68, November-December, 1992, 27; contributed by N. Vasilyev.

13. Problem M71, January-February, 1993, 25; contributed by Y. Ionin.
14. Problem M71, March-April, 1993, 24; contributed by N. Vasilyev.
15. Problem B81, May-June, 1993, 15; contributed by V. Proizvolov.
16. Problem M81, May-June, 1993, 16; contributed by S. Fomin.
17. Problem B91, September-October, 1993, 19; contributed by V. Proizvolov.
18. Problem B94, September-October, 1993, 19; contributed anonymously.
19. Problem M93, September-October, 1993, 27; contributed by S. Fomin.
20. Problem B105, January-February, 1994, 11; contributed by V. Proizvolov.
21. Problem B112, May-June, 1994, 15; contributed by A. Savin.
22. Problem B115, May-June, 1994, 15; contributed anonymously.
23. Problem M112, May-June, 1994, 19; contributed by V. Dubrovsky.
24. Problem B118, July-August, 1994, 9; contributed by A. Zilberman.
25. Exercise 3 from the article "Suggestive Tilings" by Vladimir Dubrovsky, July-August, 1994, 36.
26. Equivalent to Problem B121, September-October, 1994, 17; contributed by V. Proizvolov.

References

Solutions are generally included in the journal, but it is not always clear who is responsible for a problem or a solution. Credit is given below whenever a definite association can be made.

1. January, 1990, 41; Problem B5.
2. January, 1990, 26; Problem 7 from the article "Pigeons in Every Pigeonhole", by Alexander Soifer and Edward Lozansky.
3. September-October, 1990, 19; Problem B12.

4, 5, 6. November-December, 1990, 8; respectively Problems 5 and 6, and Exercise 8 from the article "Going To Extremes," by A. L. Rosenthal.

7. November-December, 1990, 24; Problem M18, contributed by R. Freiwald.
8. November-December, 1990, 51; from the 1990 Tournament of Towns for ages 13 to 15.
9. September-October, 1991, 49; Exercise 3 from the article "Criminal Geometry" by D. V. Fomin.
10. January-February, 1992, 44; from the article "Off Into Space," by Vladimir Dubrovsky and Igor Sharygin.
11. July-August, 1992, 27; Problem M59, contributed by B. Martynov; solution by V. Dubrovsky.
12. November-December, 1992, 27; Problem M69, contributed by V. Lev.
13. March-April, 1993, 24; Problem M80, contributed by D. Fomin.
14. March-April, 1993, 24; Problem M78, contributed by V. Proizvolov.

15. May-June, 1993, 16; contributed by A. Krasnodemskaya; solution by V. Dubrovsky.
16. September-October, 1993, 35–37; from the article "Some Things Never Change," by Yuri Ionin and Lev Kurlyandchik.
17. November-December, 1993, 27; Problem M97, contributed by V. Proizvolov; solution by V. Dubrovsky.
18. March-April, 1994, 15; Problem B110, contributed by V. Proizvolov; solution by V. Dubrovsky.
19. March-April, 1994, 31; Problem M109, contributed by D. Nyamsuren (Mongolia); solution by V. Dubrovsky.

Comments on the Exercises

Exercise 7: a is the x-coordinate of the point of intersection of the graphs of $y = \cos x$ and $y = x$; similarly b is given by the graphs of $\cos x$ and $\arcsin x$, and c by the graphs of $\sin x$ and $\arccos x$.

Exercise 15: Cut Figure 49 along BC and interchange B and C to get an isosceles triangle.

Exercise 16: Set the selected coin C aside and weigh the other 1992 coins, half on each side. There are either 19 or 20 counterfeit coins on the scale, and the pointer registers their **parity**, revealing which it is.

Exercise 17: Count the edges in a certain bipartite graph in two ways.

Exercise 18: Rotate Figure 50 through an angle of $-90°$ about the center.

Exercise 22: The brilliant solution is given in Figure 53.

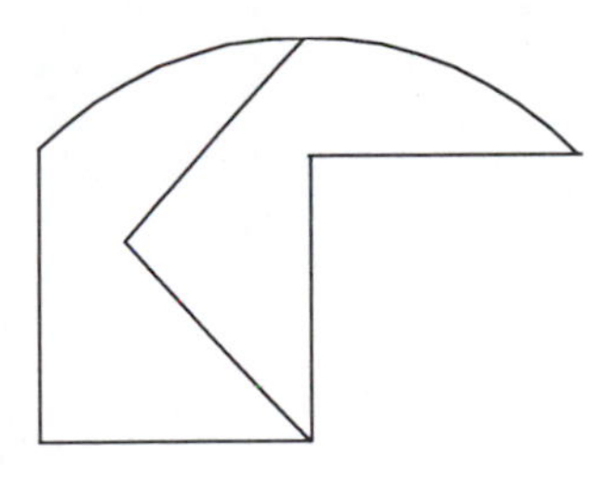

FIGURE 53

Exercise 24: The key is to determine how to get more out of your visit to the lamp-room than just seeing which lamps are on or off. To do this, throw switch 1 to heat its lamp for awhile. Then switch 1 off and put 2 on, and go into the lamp room right away. Then the lamp that is on goes with switch 2, and the hot one of the other two belongs to switch 1, the cold one to switch 3.

About *Quantum*

Quantum is a magazine about Mathematics and Science that is directed especially at secondary school and undergraduate university students. It is published every two months by the National Science Teachers Association in cooperation with the Quantum Bureau of the Russian Academy of Sciences. The 2000 personal subscription rate is $25 U.S. ($18 for students), including postage and handling in North America.

Throughout its five year existence *Quantum* has published so many well written articles of the highest scholarship and diversity that one wonders whether they can keep it up. To date, however, it's just been one wonderful issue after another. I don't know of any publication that makes science and mathematics as attractive to students.

Subscriptions may be ordered by writing to

Quantum,
Springer-Verlag New York, Inc.,
Journal Fulfillment Services Department,
P.O. Box 2485,
Secaucus, New Jersey 07094-2485,
U.S.A.

SECTION 9

Six Bulgarian Problems for 11, 12, 13, and 14 Year-Olds

These problems are given, with brief prefatorial remarks about the Bulgarian system of encouraging young mathematical talent, in an article by Kiril Bankov in the December 1991 issue of *Mathematics Competitions*, the journal of the World Federation of National Mathematics Competitions. Each contest consisted of three problems to be done in four hours. While retaining the essence of each problem, I have taken the liberty of rewording them and of omitting a small part of one of the problems.

1. For 11-Year-Olds

   Two snails, Abe and Bert, set out together to go along the same route from X to Y at seven o'clock in the morning. Abe went at a constant 12 m/hr (meters per hour). Bert started out at 8 m/hr but after 2 hours he realized he was falling farther and farther behind Abe and he climbed onto the back of a passing tortoise, named Toby, who was on his way to Y at a constant speed of 20 m/hr. Bert and Toby soon caught up to Abe and 4 hours after they had done so they reached Y.

   What time was it when Bert and Toby arrived at Y?
   How far is it from X to Y? When did Abe arrive at Y?

It is often very convenient to chart a journey of this kind on a graph by plotting "distance advanced" along the y-axis against "time" on the x-axis. Thus, in Figure 1, the segment OP from the origin O to the point $P(x, y)$ represents an advance of y units in a time of x units and the slope $\frac{y}{x}$ of OP is the speed over the section OQ of the journey.

Going at a constant speed all the way, Abe's walk is represented in Figure 2 by the segment OP of slope 12 (clearly the figure is not drawn to scale).

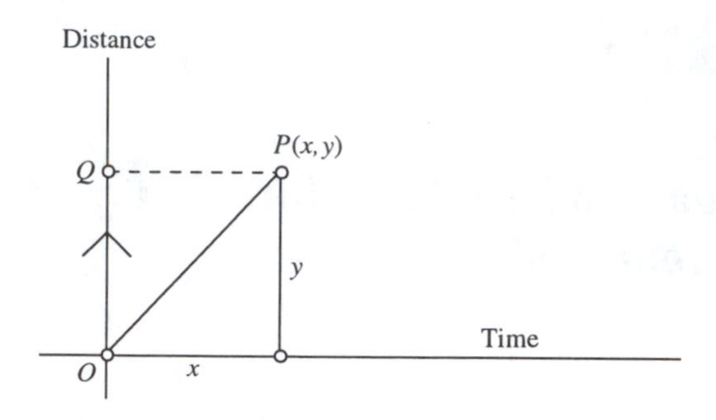

FIGURE 1

On the other hand, at 8 m/hr Bert covers 16 meters in the first 2 hours (i.e., by 9 o'clock), represented by OS (of slope 8) after which he goes the rest of the way to Y at 20 m/hr on the back of Toby, represented by SU (of slope 20).

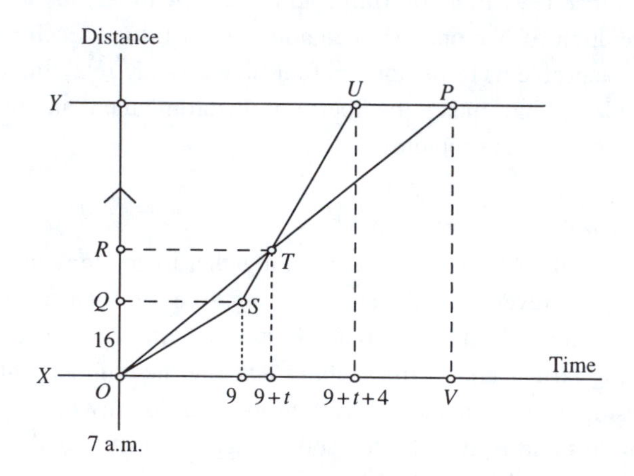

FIGURE 2

Let SU cross OP at T, at which point both Abe and Bert have advanced the same distance OR towards Y, and let the time-coordinate of T be $9 + t$ (i.e., suppose it took Bert and Toby t hours to catch Abe). Since speed $\times$ time $=$ distance, the distance OR can be calculated in two ways to yield the equation

$$12(2 + t) = 8(2) + 20(t), \quad \text{giving} \quad 8 = 8t \text{ and } t = 1.$$

Therefore Bert and Toby caught up with Abe at 10 o'clock and 4 hours later arrived at Y, i.e., at 2 p.m. Since Bert travelled 16 m on his own and $20(5) = 100$ m in the 5 hours he was on Toby's back, the distance $XY = 116$ meters.

Finally, at 12 m/hr, it took Abe $\frac{116}{12} = 9\frac{2}{3}$ hrs for the trip, and he arrived at Y at 4:40 in the afternoon.

2. For 12-Year-Olds

Peter was camping at the foot of a mountain and left camp at 10 o'clock one morning to walk to the summit. The path was horizontal for a distance and then rose to the summit. He didn't spend any time at the top but turned right around and returned to camp by the same route, arriving back at 4 p.m. If Peter walked at 4 km/hr (kilometers per hour) on level ground, ascended at 3 km/hr, and descended at 6 km/hr, how far did Peter walk altogether? Determine, within a half-hour, the time when Peter reached the summit.

A graph of Peter's trip is given in Figure 3, where he is represented by OP as having taken a hours to traverse the level section of the path (at 4 km/hr), by PQ as having taken $2b$ hours to climb to the top (at 3 km/hr), and since he descends at twice the rate at which he ascends, b hours to return to the level road (QR), and another a hours to get back to camp (RS).

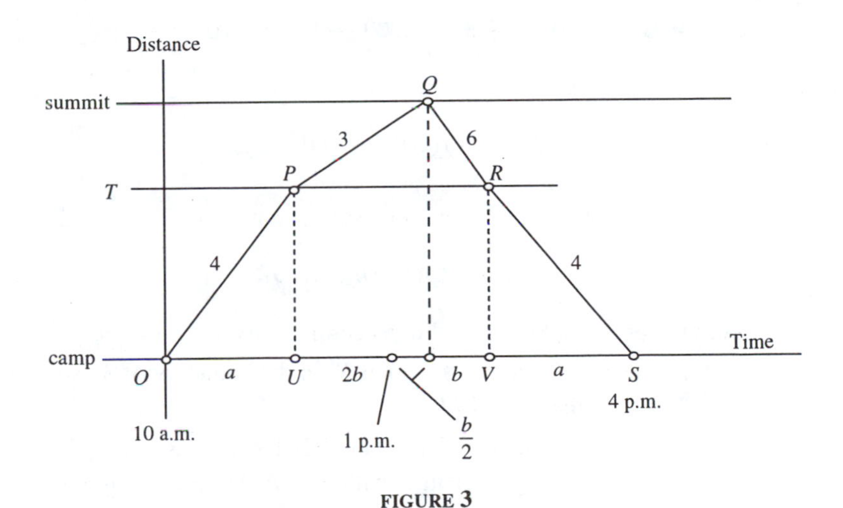

FIGURE 3

The distance from camp to the summit, then, is given by $4a + 3(2b) = 4a + 6b$. Since Peter walked from 10 a.m. to 4 p.m., a total of six hours, we have along the time axis that $2a + 3b = 6$, implying Peter walked altogether

$$2(4a + 6b) = 4(2a + 3b) = 4 \cdot 6 = 24 \text{ kilometers.}$$

Since it took Peter longer to reach the summit than to return to camp, he still had some distance to climb when half the time had elapsed, i.e., at one o'clock. Since $OU = VS$ on the time axis, one o'clock is given by the midpoint of UV, whose length is $3b$. Thus one o'clock is $\frac{3b}{2}$ from each of U and V, implying that Peter reached the summit $\frac{b}{2}$ hours after one o'clock.

Now, since $2a + 3b = 6$, it follows that $3b < 6$ and $\frac{b}{2} < 1$. Hence Peter was at the top sometime between one and two o'clock, and so, within a half-hour, he was there at one- thirty.

3. The Three Problems For 13-Year-Olds

(i) The n-digit number $A = 1599\ldots9984$, $n \geq 4$, contains $(n-4)$ 9's. B is the number obtained from A by writing it backwards. Prove that the product AB is always a perfect square.

Observe that

$$A = 1599\ldots9984 = 1600\ldots0000 - 16 = 16(10^{n-2} - 1),$$

and

$$B = 4899\ldots9951 = 4900\ldots0000 - 49 = 49(10^{n-2} - 1);$$

hence

$$AB = [4 \cdot 7(10^{n-2} - 1)]^2.$$

(ii) In Figure 4, $OA = 5$ and each other side of polygon

$$M = OABCDEFGHIJK$$

is of unit length and is parallel to a coordinate axis.

(a) Find the coordinates of the point X on the boundary of M so that OX bisects the area of M.

(b) Compare the volumes of a pyramid Π having base $\triangle OCX$ and altitude h and a right prism P with base $\triangle ODX$ and height h.

(a) A direct count of the unit squares in M shows its area is 9. Thus OX must have $4\frac{1}{2}$ units of M on each side of it.

Now, letting the point $(4, 0)$ be L (Figure 5), the area of $\triangle OLC = \frac{1}{2} \cdot 4 \cdot 1 = 2$, implying $OABC = 3$. Also, $\triangle OCD = \frac{1}{2} \cdot 1 \cdot 4 = 2$, making $OABCD = 5$. Thus OX must lie on CD such that $\triangle OCX = \frac{3}{2}$ (and

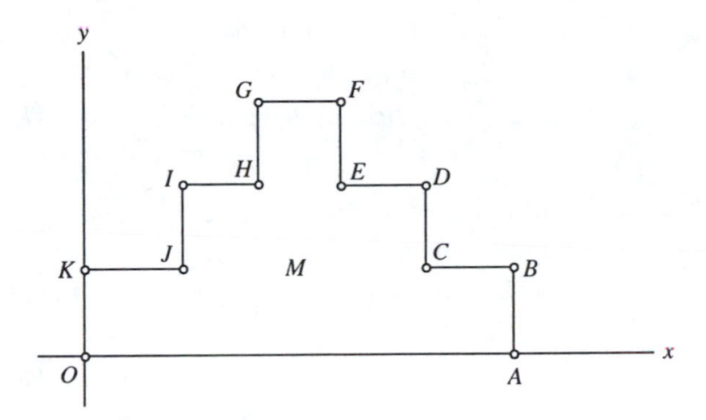

FIGURE 4

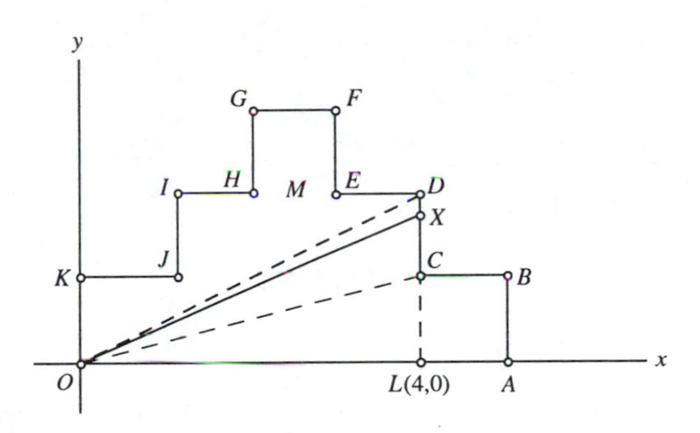

FIGURE 5

$\triangle ODX = \frac{1}{2} = \frac{1}{3} \cdot \triangle OCX$), placing X three-quarters of the way from C to D, i.e., at $(4, \frac{7}{4})$.

(b) Now, the volume of $\Pi = \frac{1}{3}$(area of base)(altitude) $= \frac{1}{3} \cdot \triangle OCX \cdot h$ and the volume of $P =$ (area of base)(height) $= \triangle ODX \cdot h = \frac{1}{3} \cdot \triangle OCX \cdot h$, and so Π and P have the same volume.

(iii) Vidin is a Bulgarian town on the Danube. A ship and a raft set off downstream from Vidin at the same time. After steaming 96 km downstream, the ship turned right around and returned to Vidin, arriving 14 hours after it had left Vidin. On the return journey the ship met the raft, floating all the while downstream, at a point 24 km from Vidin. If the ship went with its engines full out for the entire journey, that is, at its maximum stillwater speed of s km/hr,

(a) determine s and the speed v of the river,

(b) how far downstream at maximum speed would the ship be able to venture and still return to Vidin in a total of 15 hours?

(a) These straightforward journeys are charted in Figure 6, where the time taken for the ship to go the 96 km downstream is a hours and the time for the return trip is $4b$ hours. Since the ship meets the raft with only 24 km to go, that is, one-quarter of the 96 km return trip, the time taken to complete the journey is b hours (PQ), the first $3b$ hours of the return trip having been spent in returning the 72 km (NP) to the point of meeting the raft.

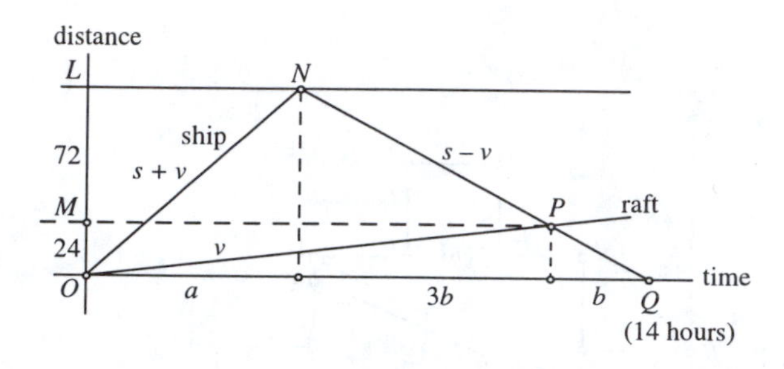

FIGURE 6

Clearly the ship proceeds downstream at the rate $s + v$, upstream at $s - v$, and the raft floats downstream at speed v. Thus we have the following system of four equations in four unknowns:

$$(s + v) \cdot a = 96,$$
$$(s - v) \cdot 4b = 96,$$
$$v \cdot (a + 3b) = 24,$$
$$a + 4b = 14.$$

A straightforward solution yields $a = 6$, $b = 2$, $v = 2$, and $s = 14$. Thus the maximum stillwater speed of the ship is 14 km/hr and the Danube flows at 2 km/hr.

(b) From part (a), it follows that the ship steams downstream at a maximum speed of $14 + 2 = 16$ and upstream at $14 - 2 = 12$, only three-quarters as quickly. Thus a return trip takes the ship $\frac{4}{3}$ as long as any downstream journey. Therefore, if the ship goes downstream a distance d in $3t$ hours, it takes

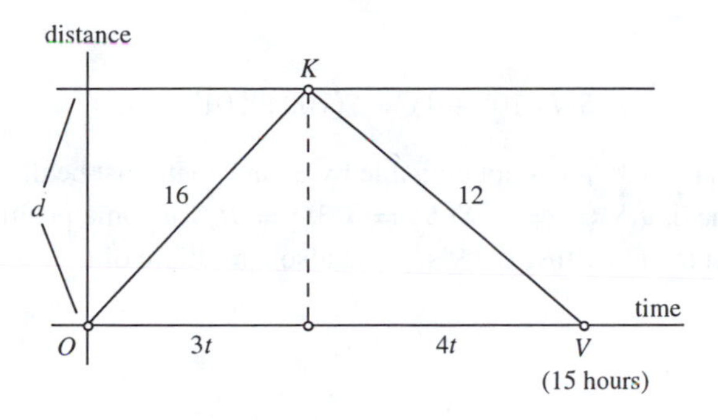

FIGURE 7

$4t$ hours to get back to Vidin, making the whole trip of duration $7t$ hours. For $7t = 15$, we have $t = \frac{15}{7}$ and the maximum distance d downstream would be

$$d = 16(3t) = 48 \cdot \frac{15}{7} = 102.8\ldots \text{ km.}$$

4. For 14-Year-Olds

A and B are positive integers in the decimal system such that (i) $A = 7B$ and (ii) the sum of the digits of A is twice the sum of the digits of B. If C is the number formed by writing the digits of B immediately after the digits of A, prove that C is a multiple of 9.

If B has k digits, then

$$\begin{aligned} C &= (\ldots \text{digits of A} \ldots)(\ldots \text{the } k \text{ digits of B} \ldots) \\ &= A \cdot 10^k + B \\ &= 7B \cdot 10^k + B \\ &= B(7 \cdot 10^k + 1). \end{aligned}$$

Let $S(n)$ denote the sum of the digits of the integer n. It is given that

$$S(A) = 2 \cdot S(B).$$

Hence

$$S(C) = S(A) + S(B) = 3 \cdot S(B),$$

and by the "rule of 3" (i.e., $3 \mid n$ if and only if $3 \mid S(n)$), it follows that $3 \mid C$, i.e.,

$$3 \mid B(7 \cdot 10^k + 1).$$

But

$$S(7 \cdot 10^k + 1) = S(700\ldots01) = 8,$$

implying that $7 \cdot 10^k + 1$ is not divisible by 3, and so it must be that $3 \mid B$. Thus $3 \mid S(B)$, making $S(C) = 3 \cdot S(B) = 3(3t) = 9t$, for some positive integer t, and by the rule of "casting out 9's," C is also a multiple of 9.

SECTION **10**

Cusumano's Challenge

The following problem was offered as a pleasant diversion to the readers of *Pi Mu Epsilon Journal* (Fall, 1994, page 25) by Andrew Cusumano of Great Neck, New York.

> The angles of equilateral triangle ABC are quartered and the arms of the angles are extended to meet as shown in Figure 1. Prove
>
> (a) EF is perpendicular to DC, and
>
> (b) GH is parallel to BC.

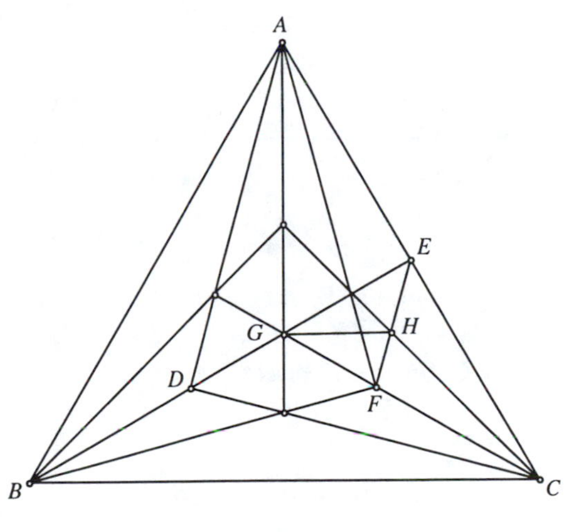

FIGURE **1**

With the problem concerning such common elements as an equilateral triangle and angle bisectors, one might expect a straightforward solution to be fairly evident. Although the problem did not prove to be particularly difficult, it resisted long enough that I was quite pleased when I saw through it.

(a) Suppose EF meets DC at X (Figure 2). Since $\triangle ABC$ is equilateral, the angle at each vertex is 60°; thus a quarter of each angle is 15°, and three quarters constitute an angle of 45°. Moreover, BE not only bisects $\angle B$, but is the altitude to AC, making $\angle AEB$ a right angle. However, in $\triangle AFB$, the 45° angles at A and B make $\angle AFB$ a right angle, too. Therefore the circle on diameter AB goes through A, E, F, and B, and $AEFB$ is a cyclic quadrilateral. Thus the exterior $\angle FEC$ at E is equal to the 45° interior $\angle ABF$ at the opposite vertex B. Finally, in $\triangle EXC$, the 45° angles at E and C make the angle at X a right angle, and hence EF is perpendicular to DC.

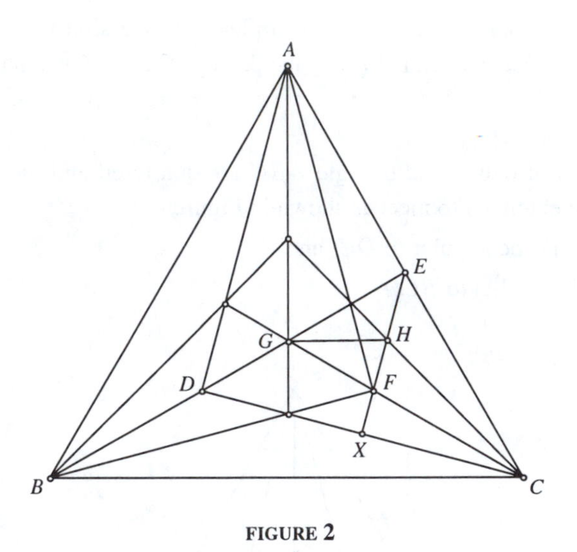

FIGURE 2

(b) In $\triangle ECG$, the angles at E and C are 90° and 30°, respectively, implying $\angle EGC = 60°$; also, since $\angle CEF = 45°$ (established in part (a)), EH is the bisector of $\angle GEC$. Since CH bisects $\angle ECG$, then H is the incenter of $\triangle ECG$ and it follows that HG is the bisector of the 60° angle $\angle EGC$. Thus

$$\angle EGH = 30° = \angle GBC,$$

and GH is parallel to BC (by equal corresponding angles).

Here is another engaging problem of Andrew Cusumano—parts (a) and (b) of Problem 817, *Pi Mu Epsilon Journal*, Fall 1994, 72; solved by two 11-year-old boys, Sammy Yu and Jimmy Yu, who are special students at the University of South Dakota, Vermillion, South Dakota.

Squares $ABDE$ and $ACFG$ are drawn outwardly on the sides of an arbitrary $\triangle ABC$ (Figure 3). BG crosses AC at Q and CE crosses AB at P; also, BG and CE intersect at R.

Prove

(a) that D, R, and F are collinear,

(b) that, with A joined to R, each of the six angles at R, other than $\angle BRC$, is $45°$.

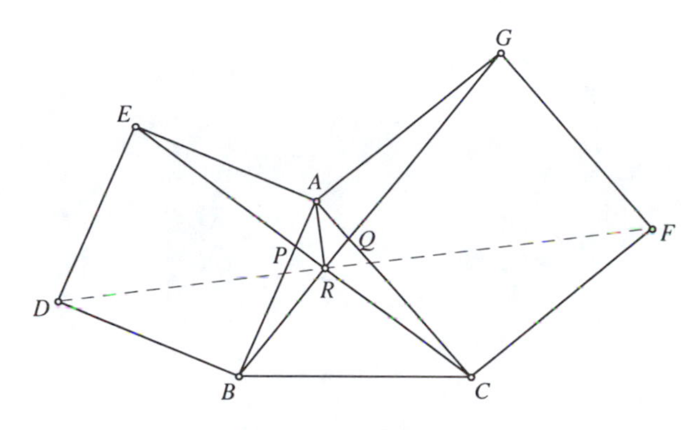

FIGURE 3

Clearly, a quarter-turn about A carries $\triangle EAC$ to $\triangle BAG$, implying EC and BG are perpendicular. Thus EB subtends a right angle at both A and R, revealing that the circle around the square $ABDE$ goes through R. Since a side of a square subtends a $90°$ angle at the center of its circumcircle, each of the angles DRB, ERD, and ARE at the circumference is $45°$. Similarly for the other three angles in question, and since $\angle DRF$ is made up of four of these $45°$ angles, it follows that D, R, and F lie in a straight line.

The Yu boys also solved the more difficult part (c) of this problem, which is left as a challenge.

(c) If $\angle ACB$ is a right angle, prove that P, Q, and F are collinear.

SECTION **11**

Five Easy Problems from the 1984 Leningrad Olympiad

1. What is the greatest number of integers between 1 and 100 inclusive that can be selected so that the sum of any two of them is divisible by 6?

There are 16 multiples of 6 up to 100, namely $\{6, 12, 18, 24, \ldots, 90, 96\}$, and it seems unlikely that one can do better than selecting these integers. In any case, observe that, for the selected set S to contain more than two members, all the integers in S must be congruent to each other (mod 6): if a and b are not congruent (mod 6), then one of

$$a + c \equiv 0 \pmod 6, \quad b + c \equiv 0 \pmod 6, \quad \text{must fail.}$$

Thus, for all $a, b \in S$, we require both

$$a \equiv b \pmod 6 \quad \text{and} \quad a + b \equiv 0 \pmod 6,$$

which rules out

$$a \text{ or } b \equiv 1,\ 2,\ 4,\ \text{and } 5 \pmod 6;$$

each member of S must be congruent either to 0 or 3 (mod 6).

Therefore, besides the 16 multiples of 6 (each of which is $\equiv 0$), the 17-member set

$$S = \{3, 9, 15, 21, \ldots, 93, 99\},$$

where each is $\equiv 3$, is also acceptable, and so the answer to the problem is not 16 but 17.

2. Find all values of x in the interval $0 < x < \pi$ such that

$$\log_{\pi/4} x + \cot x = 2.$$

Clearly $x = \frac{\pi}{4}$ is one value, for $\log_{\pi/4} \frac{\pi}{4} + \cot \frac{\pi}{4} = 1 + 1 = 2$. Now let

$$f(x) = \log_{\pi/4} x + \cot x - 2.$$

Then

$$\frac{d[f(x)]}{dx} = \frac{1}{x \ln \frac{\pi}{4}} - \cot^2 x - 1.$$

Since $\frac{\pi}{4} < 1$, then $\ln \frac{\pi}{4} < 0$, and for $0 < x\pi$, $\frac{d[f(x)]}{dx}$ is always negative, implying that $f(x)$ is a strictly monotonic function, in which case $f(x)$ takes each of its values only once. Hence $x = \frac{\pi}{4}$ is the only value.

3. Without calculating the numerical values of the roots, show that

$$\sqrt[3]{4} - \sqrt[3]{10} + \sqrt[3]{25} > \sqrt[3]{6} - \sqrt[3]{9} + \sqrt[3]{15}.$$

Transposing, we wish to show that

$$4^{1/3} + 9^{1/3} + 25^{1/3} > 6^{1/3} + 10^{1/3} + 15^{1/3},$$

that is,

$$(2^{1/3})^2 + (3^{1/3})^2 + (5^{1/3})^2 > 2^{1/3} \cdot 3^{1/3} + 2^{1/3} \cdot 5^{1/3} + 3^{1/3} \cdot 5^{1/3},$$

or, with $2^{1/3} = a$, $3^{1/3} = b$, $5^{1/3} = c$, that

$$a^2 + b^2 + c^2 > ab + ac + bc.$$

But this follows easily from the Cauchy inequality:

$$(a^2 + b^2 + c^2)(1^2 + 1^2 + 1^2) \geq (a \cdot 1 + b \cdot 1 + c \cdot 1)^2,$$

with equality only for $a = b = c$. Since $a \neq b$, then we have

$$\begin{aligned} 3(a^2 + b^2 + c^2) &> (a + b + c)^2 \\ &= a^2 + b^2 + c^2 + 2(ab + bc + ca), \end{aligned}$$

giving

$$a^2 + b^2 + c^2 > ab + bc + ca,$$

as desired.

4. Find all prime numbers a and b such that $a^b + b^a$ is also a prime.

Suppose $a^b + b^a = p$, a prime number. Since a and b are prime numbers, then each of a and b is at least 2, making $p > 2$ and hence an odd prime. This, in turn, implies one of a, b must be odd and the other even, forcing one to be the only even prime, 2. Thus for some odd prime b we have

$$2^b + b^2 = p.$$

Clearly $b = 3$ yields the solution

$$2^3 + 3^2 = 8 + 9 = 17.$$

However, if $b > 3$, then, being a prime number,

$$b \equiv 1 \text{ or } 2 \pmod 3 \quad \text{and} \quad b^2 \equiv 1 \pmod 3.$$

In this case, we have modulo 3 that

$$2^b + b^2 \equiv (-1)^b + 1 \equiv 0, \text{ since } b \text{ is odd.}$$

Thus, for $b > 3$, $2^b + b^2$ is always divisible by 3, and being greater than 3, it cannot be a prime number. Thus the only solution is $(a, b) = (2, 3)$.

5. Prove that the area of a triangle never exceeds one-sixth the sum of the squares of the lengths of its sides, i.e., for $\triangle ABC$ with sides a, b, c,

$$\triangle ABC \leq \frac{1}{6}(a^2 + b^2 + c^2).$$

If Δ is the area of $\triangle ABC$, then

$$\Delta = \frac{1}{2}ab \sin C = \frac{1}{2}bc \sin A = \frac{1}{2}ca \sin B,$$

giving

$$ab = \frac{2\Delta}{\sin C} \geq 2\Delta, \quad \text{since } \sin C \leq 1.$$

Similarly

$$bc \geq 2\Delta \quad \text{and} \quad ca \geq 2\Delta,$$

and

$$ab + bc + ca \geq 6\Delta.$$

But, as we saw in Problem 3,

$$a^2 + b^2 + c^2 \geq ab + bc + ca,$$

and so

$$a^2 + b^2 + c^2 \geq 6\Delta,$$

from which the desired result follows immediately.

SECTION 12
An Arithmetic Puzzle

This section is based on the note "On Some Mathematical Recreations," by Richard Bellman (Rand Corporation, Santa Monica, California), published in the *American Mathematical Monthly*, 1962, 640–643, and reprinted in *Selected Papers on Precalculus*, Volume 1, Raymond W. Brink Series, Mathematical Association of America, 1977, 74–76.

It is easy to check that

$$1 + 2 + 3 + 4 + 5 + 6 + 7 + 8 \times 9 = 100.$$

Among the various ways of converting

$$1 * 2 * 3 * 4 * 5 * 6 * 7 * 8 * 9 = 100$$

into a valid equation by independently changing each $*$ either to a $+$ sign or a $\times$ sign, what is the **minimum number of $+$ signs** that must be used?

1. Using the notation $f_n(k)$ for the minimum number of $+$ signs that must be used to validate the equation

$$1 * 2 * 3 * \cdots * n = k,$$

our task is to determine $f_9(100)$.

The key is to focus on the possible positions of the **last** $+$ sign. Clearly, any expression ending in $7 \times 8 \times 9$ will exceed 100, and so the last $+$ sign must occur either between the 7 and the 8 or between the 8 and the 9:

$$\text{(i)} \quad 1 * 2 * 3 * 4 * 5 * 6 * 7 + 8 \times 9 = 100,$$

$$\text{(ii)} \quad 1 * 2 * 3 * 4 * 5 * 6 * 7 * 8 + 9 = 100.$$

Case (i): Since $8 \times 9 = 72$, case (i) requires $1 * 2 * 3 * 4 * 5 * 6 * 7 = 28$. Now, the sum of all seven of these digits is 28, and it is pretty clear that for $k \geq 3$ a smaller contribution to the value of the expression is obtained by adding k than by multiplying by k. Thus

$$2 * 3 * 4 * 5 * 6 * 7 \geq 2 + 3 + 4 + 5 + 6 + 7 = 27,$$

implying the only way to get as little as 28 is by using + signs all the way. Hence the only solution in case (i) is $1 + 2 + 3 + 4 + 5 + 6 + 7 + 8 \times 9 = 100$, containing seven + signs.

Case (ii): In this case, we must have $1 * 2 * 3 * 4 * 5 * 6 * 7 * 8 = 91$, for a total of at least $1 + f_8(91)$ + signs in the complete expression for 100. Thus

$$f_9(100) = \min\{7, 1 + f_8(91)\}. \tag{1}$$

2. Treating $f_8(91)$ in the same way, we have $6 \times 7 \times 8 > 91$, and from

$$1 * 2 * 3 * 4 * 5 * 6 * 7 + 8 = 91 \quad \text{and} \quad 1 * 2 * 3 * 4 * 5 * 6 + 7 \times 8 = 91,$$

we obtain

$$f_8(91) = \min\{1 + f_7(83), 1 + f_6(35)\}. \tag{2}$$

Carrying on with $f_7(83)$ and $f_6(35)$, we have,

$$\text{since } 5 \times 6 \times 7 > 83, \text{ that } f_7(83) = \min\{1 + f_6(76), 1 + f_5(41)\},$$
$$\text{and since } 4 \times 5 \times 6 > 35, \text{ that } f_6(35) = \min\{1 + f_5(29), 1 + f_4(5)\}.$$

Now, clearly there is no way to satisfy $1 * 2 * 3 * 4 = 5$, and therefore

$$f_6(35) = 1 + f_5(29),$$

and since it is easy to see that $f_5(29) = 1$, it follows that $f_6(35) = 2$:

because $1 \times 2 \times 3 \times 4 \times 5 = 120 > 35$, at least one + sign must be used, and $1 \times 2 \times 3 \times 4 + 5 = 29$ demonstrates that one is sufficient.

Thus we have from (2) that

$$f_8(91) = \min\{1 + f_7(83), 1 + f_6(35)\} = \min\{1 + f_7(83), 3\}. \tag{3}$$

It remains to finish dealing with $f_7(83)$, which we shall discover is another dead-end. We have observed that

$$f_7(83) = \min\{1 + f_6(76), 1 + f_5(41)\}.$$

Since $4 \times 5 \times 6 > 76$ and $3 \times 4 \times 5 > 41$, we have

$$f_6(76) = \min\{1 + f_5(70), 1 + f_4(46)\},$$

in which both alternatives are unrealizable:

> clearly it is beyond the capabilities of $1 * 2 * 3 * 4$ to produce 46, and with regard to $1 * 2 * 3 * 4 * 5$ yielding 70, we have the following: while $2 \times 3 \times 4 \times 5 > 70$, it is not true that $3 \times 4 \times 5 > 70$, and we need to consider the three cases
>
> $$\begin{aligned} 1 * 2 * 3 * 4 + 5 &= 70, \text{ requiring the impossible } 1 * 2 * 3 * 4 = 65, \\ 1 * 2 * 3 + 4 \times 5 &= 70, \text{ requiring the impossible } 1 * 2 * 3 = 50, \\ 1 * 2 + 3 \times 4 \times 5 &= 70, \text{ requiring the impossible } 1 * 2 = 10. \end{aligned}$$

It follows, then, that there is no way to generate 83 from $1*2*3*4*5*6*7$. Hence $1 + f_7(83)$ is not an option in (3) and we have $f_8(91) = 3$.

Finally, (1) gives

$$f_9(100) = \min\{7, 1 + f_8(91)\} = \min\{7, 1 + 3\} = 4,$$

which is realized in the equation

$$1 \times 2 \times 3 \times 4 + 5 + 6 + 7 \times 8 + 9 = 100.$$

SECTION 13

A Few Gleanings from *The Mathematical Gazette*

The engaging problems in this section were gleaned from various issues of the outstanding British journal *The Mathematical Gazette*. The references are to this journal.

1. (1936, **95**)

   In Figure 1, TA is a tangent to a given circle. Prove that, for all choices of secant TBC, the bisector of $\angle T$ cuts an **isosceles** triangle AXY from $\triangle ABC$.

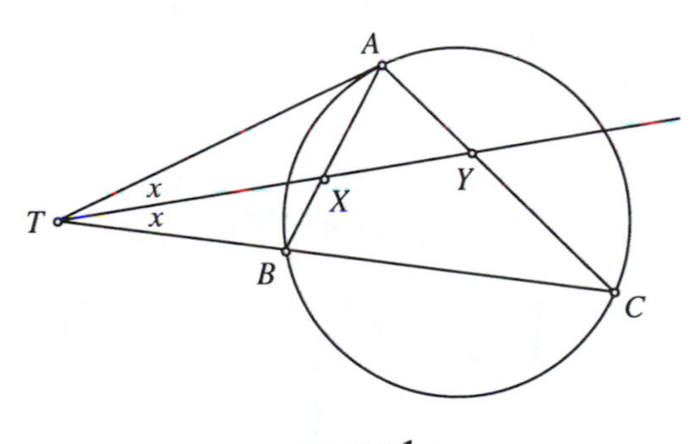

FIGURE 1

Since AT is a tangent and AB a chord, we have

$$\angle TAB = \angle C,$$

and since TXY bisects $\angle T$, then triangles TAX and TYC are equiangular, and their third angles TXA and TYC are equal. Thus the supplements of these third angles are also equal, namely $\angle AXY$ and $\angle AYX$, and the conclusion follows.

Problems 2 and 3 are sample problems that were included in a review of the splendid book *Mathematical Puzzling*, by Tony Gardiner, University of Birmingham (1988, **145**).

2. In Figure 2, three circles of radius 5 are arranged side by side in a row. From the point A on the first circle which is in line with their centers, a tangent AD is drawn to the third circle. What is the length of the intersection BC of this tangent with the middle circle?

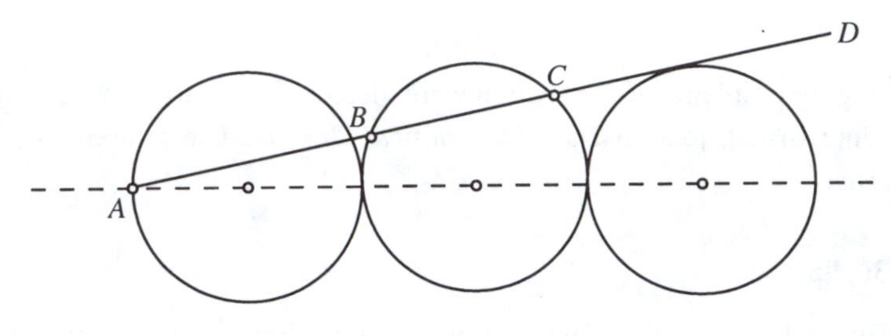

FIGURE 2

Let the centers of the second and third circles be E and F (Figure 3) and let perpendiculars from E and F meet the tangent at G and H. Then H is the point of contact of the tangent and G bisects the chord BC.

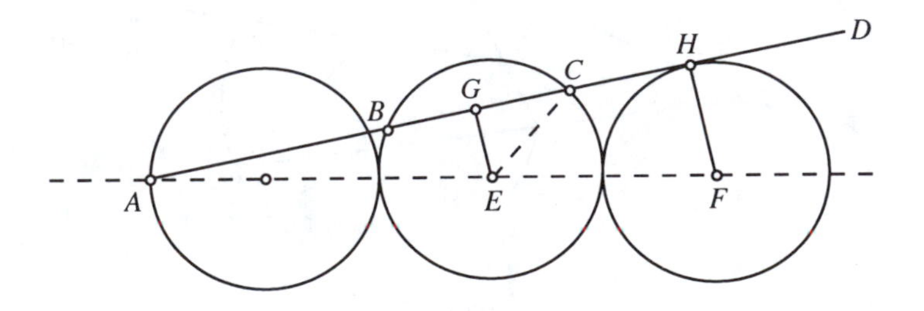

FIGURE 3

Clearly the triangles AEG and AFH are similar, giving

$$\frac{EG}{HF} = \frac{AE}{AF} = \frac{15}{25} = \frac{3}{5},$$

and since the radius $FH = 5$, then $EG = 3$. But EC is also the radius 5, and therefore $\triangle ECG$ is a 3-4-5 right triangle, making $GC = 4$ and $BC = 8$.

3. Five circles are arranged in a row so that each touches the circle(s) next to it and the straight lines AB and CD (Figure 4). If the radius of the first circle is 12 and the last circle is 18, what is the radius of the middle circle?

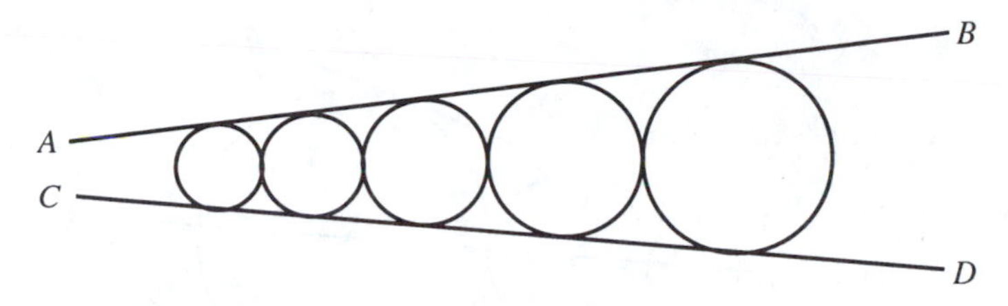

FIGURE 4

Clearly AB and CD are not parallel, since the circles are not all the same size; let AB and CD meet at O. First, consider any pair of consecutive circles in the row (Figure 5); suppose their centers are P and Q, their radii a and b, that R and S are the points of contact with AB, and let $\angle O = 2\alpha$. Then the line of centers OPQ bisects $\angle O$, and if the perpendicular from P to QS meets it at T, we have

$$\begin{aligned}&RPTS \text{ is a rectangle,}\\ &QT = QS - TS = b - a,\\ &PQ = b + a,\end{aligned}$$

and since PT is parallel to OB,

$$\angle TPQ = \angle BOQ = \alpha.$$

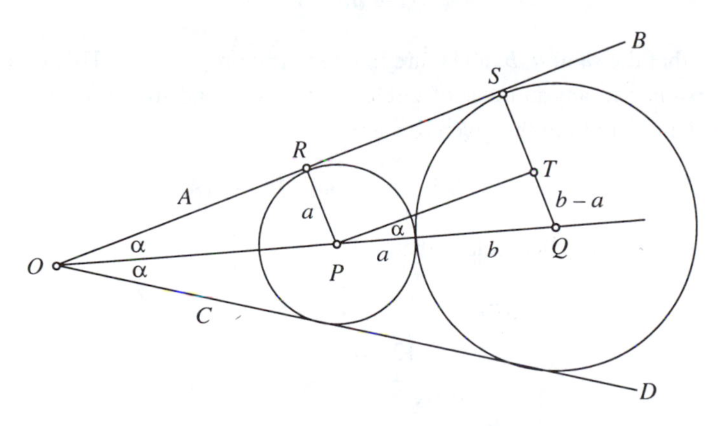

FIGURE 5

Hence, from right triangle PQT,

$$\sin\alpha = \frac{TQ}{PQ} = \frac{b-a}{b+a}.$$

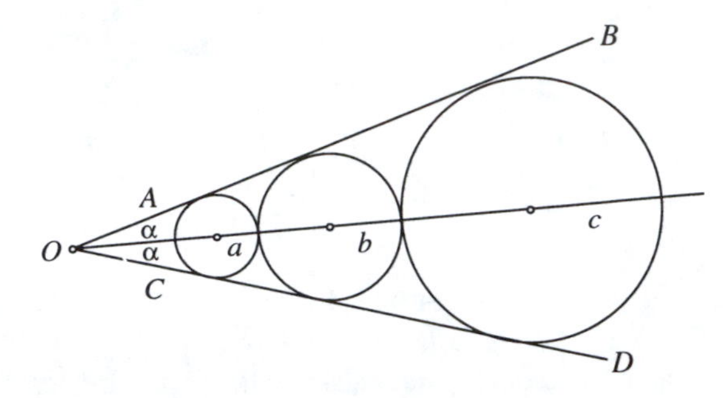

FIGURE 6

A row of three such circles, of radii a, b, c (Figure 6), would yield

$$\sin\alpha = \frac{b-a}{b+a} = \frac{c-b}{c+b},$$

giving

$$bc + b^2 - ac - ab = bc - b^2 + ac - ab,$$
$$2b^2 = 2ac$$

and

$$b^2 = ac,$$

implying that the radii a, b, and c are in geometric progression. This property clearly extends to any number of circles in the row, and therefore, for some value r, the five radii in the given row are

$$12,\ 12r,\ 12r^2,\ 12r^3,\ \text{and}\ 12r^4 = 18.$$

The radius of the middle circle is then

$$\begin{aligned} 12r^2 &= \sqrt{12 \cdot 12r^4} \\ &= \sqrt{12 \cdot 18} \\ &= 6\sqrt{6} \\ &= 14.7 \text{ approximately.} \end{aligned}$$

4. (This problem was posed by A. P. Rollett in the note “A Curious Rectangle”; 1937, **412**)

   In a circle K with center A, circles with half the radius of K are drawn with centers B and E to touch K and to touch each other at A (Figure 7). A circle with center D is drawn to touch these three circles, and finally a circle with center C is drawn to touch the three circles in its vicinity. Prove that the centers A, B, C, D determine a rectangle.

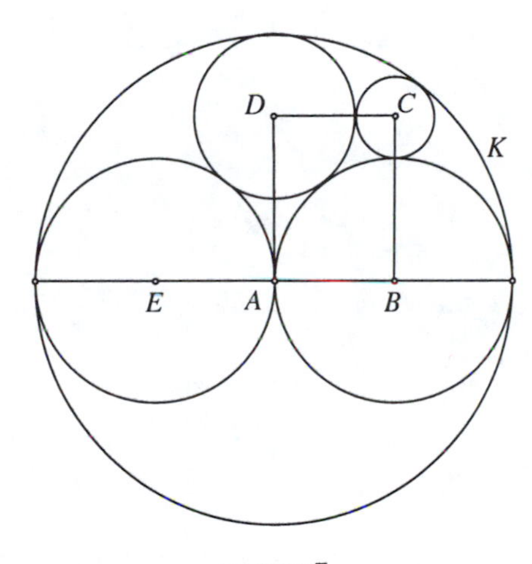

FIGURE 7

Instead of constructing the final circle, complete the rectangle $DABP$ (Figure 8), with DP crossing the circle with center D at Q and the circle with center B at R.

Let radius $AB = a$, radius $DQ = r$, and let AD meet K at S. Then, since the line joining the centers of touching circles goes through their point of contact, S is the point of contact of K and the circle with center D, making AS and DS radii of these circles, and we have

$$AS = 2a = AD + r, \quad \text{giving} \quad PB = AD = 2a - r.$$

Therefore

$$PR = PB - RB = (2a - r) - a = a - r,$$

and

$$PQ = DP - DQ = AB - DQ = a - r.$$

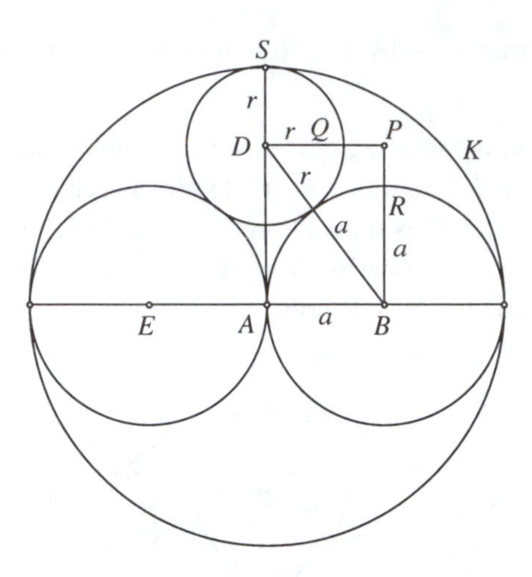

FIGURE 8

Hence the circle with center P and radius $a-r$ touches the circles with centers D and B at Q and R.

But the distance between the centers A and P is the diagonal of the rectangle, which is also given by the other diagonal $DB = r + a$, and since

$$r + a = 2a - (a - r),$$

the distance AP between the centers is equal to the difference of the radii, implying that K and the circle with center P touch internally. Thus the circle with center P and radius $a - r$ touches the three circles in its vicinity and it follows that P is in fact the center C. The conclusion follows.

5. (Based on the note by J. V. Narlikar, University College, Cardiff, 1981, **32**, and the solution by E. H. Lockwood, Charlminster, Dorset (the author of the wonderful volume *A Book of Curves*, Cambridge University Press, 1961).)

   E, F, and G are the vertices of an equilateral triangle inscribed in a circle center O. A 3-petal lotus, shaded in Figure 9, is determined by circles with centers E, F, G that pass through O. Prove that the total unshaded area in the circle $= 2\triangle EFG$.

Let S be the area of the given circle and u the area of the segment AFC (Figure 10). Since reflection in AC takes segment AFC to AOC, we have

$$2u = \frac{1}{3}(\text{total unshaded area}) + 2 \text{ petals},$$

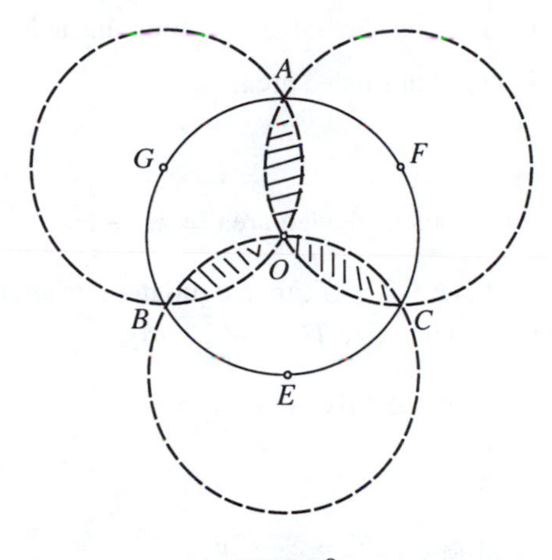

FIGURE 9

giving

$$6u = \text{total unshaded area} + 6 \text{ petals}.$$

But clearly

$$3 \text{ petals} = S - \text{total unshaded area},$$

and therefore

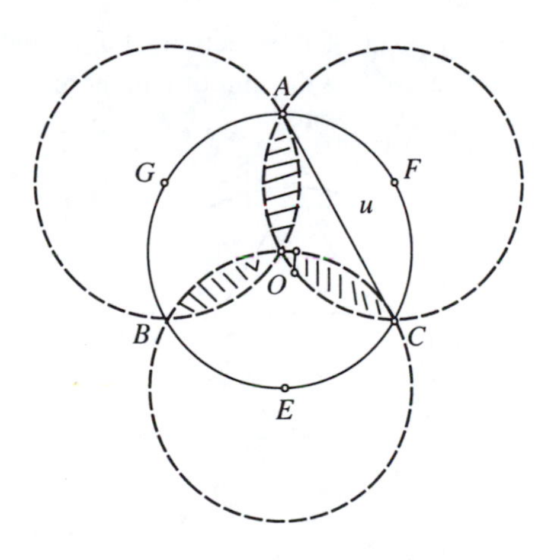

FIGURE 10

$$\begin{aligned} 6u &= \text{total unshaded area} + 2(S - \text{total unshaded area}) \\ &= 2S - \text{total unshaded area.} \end{aligned}$$

Thus

$$\text{the total unshaded area} = 2S - 6u.$$

Letting the area of $\triangle EFG = T$, then equilateral triangle ABC, which is congruent to $\triangle EFG$, also has area T. Now, clearly

$$\triangle ABC + 3u = S,$$

giving

$$T = S - 3u,$$

and hence the total unshaded area $= 2S - 6u = 2T = 2\triangle EFG$.

(J. V. Narlikar pointed out the interesting fact that the total area of these unshaded circular regions is given by a value that does not involve the number π and Lockwood took the idea even further by showing that the area can be determined by an argument that nowhere refers to π.)

6. (1988, **29**; solution by John Rigby, University College, Cardiff)

   L and M are midpoints of sides of equilateral triangle ABC, and LM meets the circumcircle of the triangle at Y (Figure 11). Prove that $\frac{LM}{MY}$ is the golden ratio.

Let X be the second point of intersection of LM and the circle (Figure 12), let $LM = x$ and let MY be the unit of length. Since L and M are the

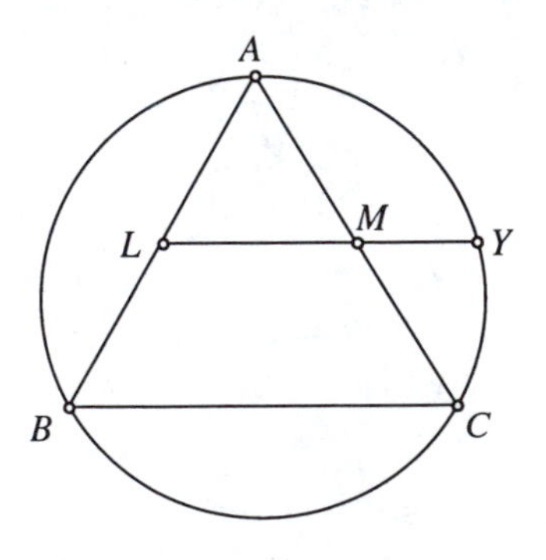

FIGURE 11

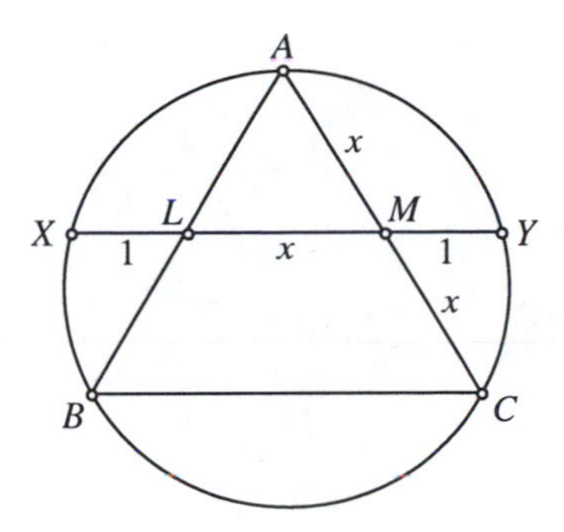

FIGURE 12

midpoints of AB and AC, we have

$$x = LM = \frac{1}{2}BC = \frac{1}{2}AC = AM = MC,$$

and since $\triangle ABC$ is equilateral, $XL = MY = 1$. Then

$$XM \cdot MY = AM \cdot MC$$

gives

$$(1 + x) \cdot 1 = x \cdot x,$$
$$x^2 - x - 1 = 0,$$

and since $x > 0$,

$$x = \frac{1 + \sqrt{5}}{2}, \text{ the golden ratio.}$$

7. (This is the converse of the problem given in Note 70.17, by J. R. Groggins, Glasgow, Scotland, 1986, **133**)

   In $\triangle ABC$, suppose BE and CF each bisects the perimeter (Figure 13). If BE and CF cross at N, prove that AN extended also bisects the perimeter.

As usual, let the lengths of the sides of $\triangle ABC$ be a, b, c, and its semiperimeter be s. Then, referring to Figure 14, since BE bisects the perimeter,

$$s = BA + AE = c + AE,$$

giving

$$AE = s - c;$$

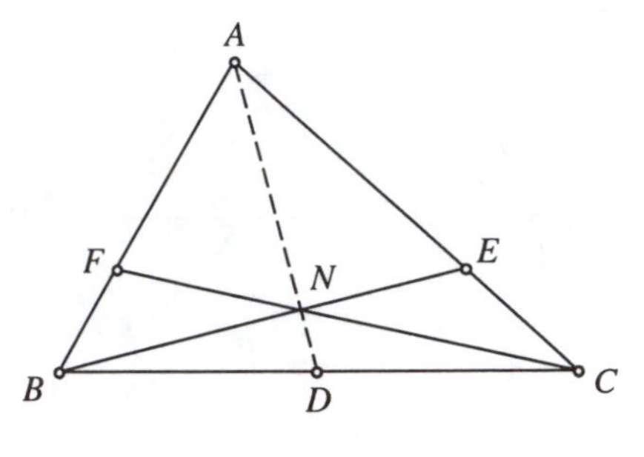

FIGURE 13

similarly, from the bisector CF,

$$AF = s - b.$$

Also,

$$s = BC + BF = BC + CE,$$

giving

$$BF = CE = s - BC = s - a.$$

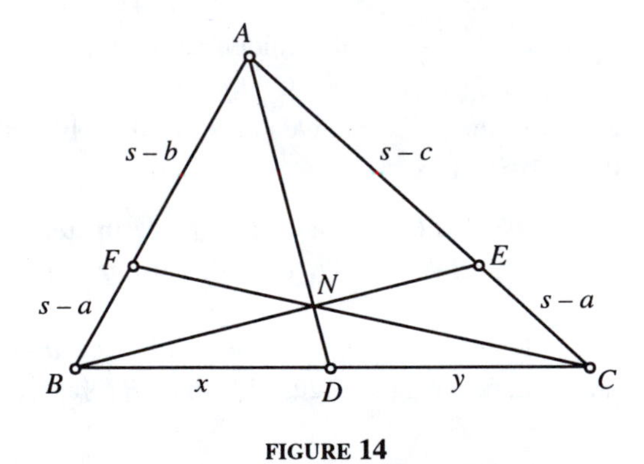

FIGURE 14

Let AN meet BC at D, and let $BD = x$ and $DC = y$. Then, by Ceva's theorem,

$$\frac{s-b}{s-a}\cdot\frac{x}{y}\cdot\frac{s-a}{s-c}=1,$$
$$\frac{s-b}{s-c}=\frac{y}{x},$$
$$\frac{s-b}{s-c}+1=\frac{y}{x}+1,$$
$$\frac{s-b+s-c}{s-c}=\frac{y+x}{x},$$
$$\frac{2s-b-c}{s-c}=\frac{a}{s-c}=\frac{y+x}{x}=\frac{a}{x},$$

giving

$$x=s-c.$$

Then

$$AB+BD=c+x=s,$$

implying AD also bisects the perimeter.

8. (From Note 2144, by E. J. Hopkins, 1950, 130–131; the solution is due to L. C. Lyness)

   In $\triangle ABC$, let $\angle A$ be reduced by closing in each of the sides AB and AC through an angle α; similarly let angles B and C be reduced by angles of 2β and 2γ, to determine the three points X, Y, and Z in $\triangle ABC$ as in Figure 15.

   Prove that, for all choices of α, β, and γ, the lines AX, BY, and CZ are concurrent.

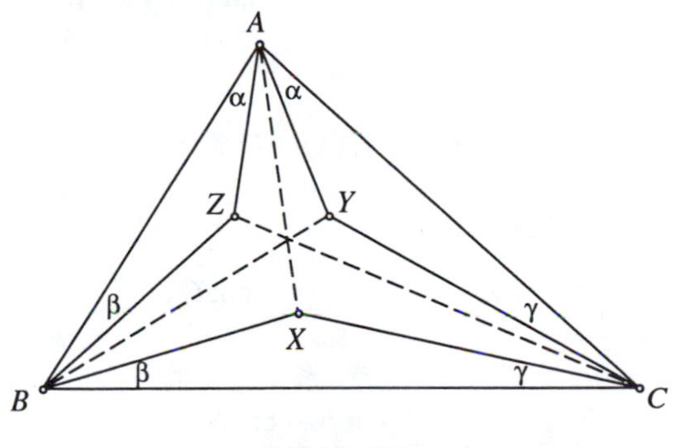

FIGURE 15

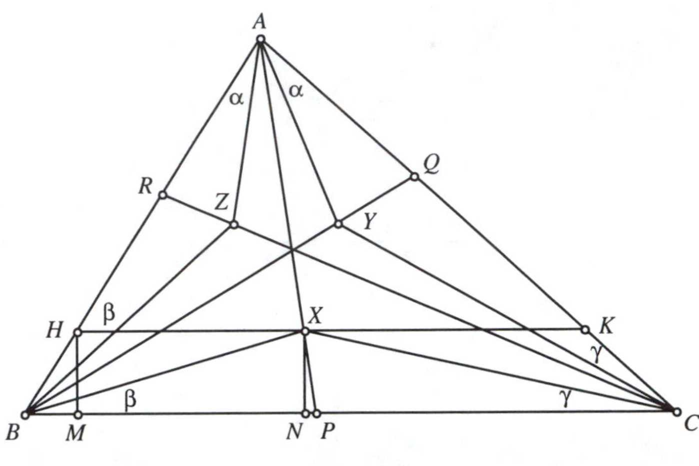

FIGURE 16

Let AX, BY, CZ meet the opposite sides at P, Q, R (Figure 16). Let the line through X which is parallel to BC meet AB at H and AC at K. Finally, let perpendiculars to BC from H and X meet BC at M and N to give rectangle $HMNX$.

Since HXK is parallel to BC, we have, from the pairs of similar triangles (AHX, ABP) and (AXK, APC), that

$$\frac{BP}{PC} = \frac{HX}{XK}.$$

Thus the ratio $\frac{BP}{PC}$ can be determined by finding HX and XK.

Now, from rectangle $HMNX$, we have

$$\begin{aligned} HX = MN &= BN - BM \\ &= XN \cot\beta - HM \cot B \quad \text{(from triangles } BXN \text{ and } BHM\text{)}, \end{aligned}$$

and since $HM = XN$, then

$$HX = XN(\cot\beta - \cot B).$$

Similarly,

$$XK = XN(\cot\gamma - \cot C),$$

and we obtain

$$\frac{BP}{PC} = \frac{\cot\beta - \cot B}{\cot\gamma - \cot C}.$$

In the same way,

$$\frac{CQ}{QA} = \frac{\cot\gamma - \cot C}{\cot\alpha - \cot A} \quad \text{and} \quad \frac{AR}{RB} = \frac{\cot\alpha - \cot A}{\cot\beta - \cot B}.$$

Clearly, then,

$$\frac{AR}{RB} \cdot \frac{BP}{PC} \cdot \frac{CQ}{QA} = 1,$$

and AX, BY, and CZ are concurrent by Ceva's theorem.

Observe that essentially the same proof goes through if the angles of $\triangle ABC$ are enlarged by opening the arms equally instead of reducing them as in the present case.

SECTION **14**

Three Problems from the 1994 Putnam Contest

1. Suppose that $a_1, a_2, a_3, \ldots$ is a sequence of positive numbers such that, for all $n \geq 1$,

$$a_{2n} + a_{2n+1} \geq a_n.$$

Prove the series

$$S = a_1 + a_2 + a_3 + \cdots \quad \text{diverges.}$$

Grouping the terms of S "at the powers of 2," we get

$$S = (a_1) + (a_2 + a_3) + (a_4 + a_5 + a_6 + a_7) + \cdots.$$

Each bracket contains twice as many terms (2^k) as the preceding bracket (2^{k-1}) and, in fact, it consists precisely of the pairs $a_{2n} + a_{2n+1}$ for the terms a_n of the preceding bracket: for example

> $(a_4 + a_5 + a_6 + a_7)$ contains the $(a_4 + a_5)$ of the a_2 of the preceding bracket and the $(a_6 + a_7)$ of a_3 (This is easily proved in general.)

Because $a_{2n} + a_{2n+1} \geq a_n$, each bracket is at least as big as the preceding one and hence S can be written as a series of **nondecreasing** positive terms, and is therefore divergent.

2. Each point of isosceles right triangle ABC, with legs AC and BC of unit length, is colored one of four colors 1, 2, 3, 4. Prove that some two points of the triangle, whose distance apart is at least $d = 2 - \sqrt{2}$, must have the same color.

The ideal solution of this problem would be to find a set S of 4 + 1 = 5 points of the triangle such that the distance between any two of them is at least d; then, in coloring S with only four colors, some color would have to be repeated, establishing the claim. Unfortunately, I was unable to find such a set. However, there is a set of six points which is just as good.

Let us try to color the triangle without satisfying the claim; when we find it can't be done, it will have to be admitted that the claim is valid.

First, note that $d = 2 - \sqrt{2}$ is approximately .6, which is less than the length of a leg, and that the hypotenuse $AB = \sqrt{2} > 2d$. Now, let the points X, Y, Z be taken on AC, AB, and BC, respectively, so that

$$AX = AY = BZ = d \quad \text{(Figure 1).}$$

Then XZ cuts off a little isosceles right triangle CXZ at C with leg

$$1 - d = 1 - (2 - \sqrt{2}) = \sqrt{2} - 1.$$

Its hypotenuse, then, is

$$XZ = (\sqrt{2} - 1) \cdot \sqrt{2} = 2 - \sqrt{2} = d.$$

Clearly, then, the distance between any two of the points A, X, Z, B is at least d, and therefore, in order to avoid satisfying the claim, no two of them can be given the same color. Suppose they are colored 1, 2, 3, 4 as in Figure 1.

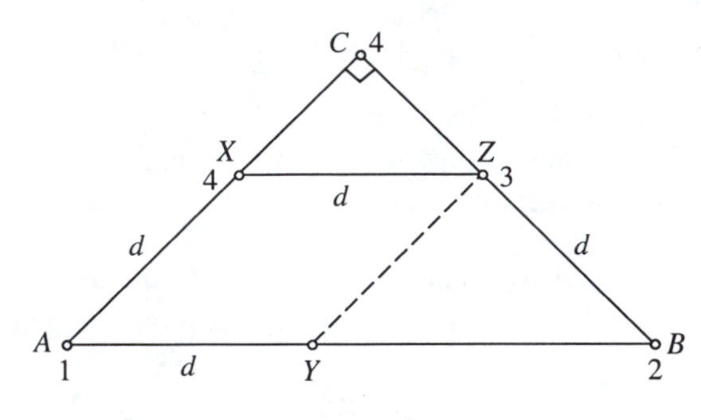

FIGURE 1

Now, the vertices of the triangle are too far apart for two of them to be the same color, and so C must be colored either 3 or 4; for definiteness, suppose C is colored 4. Similarly, recalling $AB > 2d$, Y is too far away to share a color with a vertex, leaving it only color 3, which would make it the same as Z. But, since XZ is equal and parallel to AY, $AXZY$ is a parallelogram and $YZ = AX = d$. Thus Y can't even be colored 3 without validating the claim, and the argument is complete. (If C were colored 3, then, we would take Y on BC so that $BY = d$, rather than $AY = d$, and consider YX $(= d)$ instead of YZ.)

3. Find all positive integers n that are within 250 of exactly 15 perfect squares. (Clarifications: (i) The perfect squares are $\{0, 1, 4, 9, \ldots\}$, including 0; (ii) the "within" is not strict, that is, n is within r of m if the difference between n and m is less than **or equal to** r; e.g. 300 is within 100 of 200.)

We seek all positive integers n such that the closed interval $I = [n - 250, n + 250]$ contains exactly 15 perfect squares. Of course, such squares would have to be consecutive, and so, for some positive integer $x \geq 7$, they would be the 15 squares

$$(x-7)^2, (x-6)^2, \ldots, x^2, \ldots, (x+7)^2.$$

In order for the interval I, whose length is 500, to contain these squares, it must be that

$$\begin{aligned}(x+7)^2 - (x-7)^2 &\leq 500,\\ 28x &\leq 500,\end{aligned}$$

yielding

$$x \leq 17 \quad (\text{since } 28 \cdot 18 = 504).$$

Therefore the squares in question must be 15 consecutive values from the squares up to $(17+7)^2 = 24^2$:

$$\{0, 1, 4, \ldots, 529, 576\}.$$

Now, for the interval I to contain precisely the 15 squares

$$(k+1)^2, (k+2)^2, \ldots, (k+15)^2,$$

the interval cannot extend as far as the next square on either side of them, that is, it can't reach as far as k^2 on the left nor as far as $(k+16)^2$ on the right. Hence I is contained in the interval $[k^2+1, (k+16)^2 - 1]$, and we have

$$\begin{aligned}[(k+16)^2 - 1] - (k^2+1) &\geq 500,\\ 32k + 254 &\geq 500,\\ k &\geq 8.\end{aligned}$$

Thus the smallest of the 15 squares must be at least $(k+1)^2 = 9^2 = 81$, which restricts the string of squares to the two sets

$$\text{(i) } \{81, 100, 121, \ldots, 529\} \quad \text{or} \quad \text{(ii) } \{100, 121, \ldots, 576\}.$$

For set (i) to lie in the interval $I = [n - 250, n + 250]$, we must have

$$n - 250 \geq 8^2 + 1 = 65 \quad \text{and} \quad n + 250 \leq 24^2 - 1 = 575,$$

$$n \geq 315 \quad \text{and} \quad n \leq 325,$$

that is,

$$n = 315, 316, \ldots, 325.$$

These values are all acceptable since they yield the intervals $[65, 565]$, $[66, 566], \ldots, [75, 575]$, each of which contains the set (i).

Similarly, for set (ii), $\{100, 121, \ldots, 576\}$, we must have

$$n - 250 \geq 82 \quad \text{and} \quad n + 250 \leq 624,$$

giving

$$n \geq 332 \quad \text{and} \quad n \leq 374,$$

suggesting

$$n = 332, 333, \ldots, 374.$$

However, in order to include the square 100, n cannot exceed $100+250 = 350$. Therefore this case actually yields only the values $n = 332, 333, \ldots, 350$, and since the corresponding intervals

$$[82, 582], [83, 583], \ldots, [100, 600]$$

all contain set (ii), we conclude that n may be any integer from 315 to 350, except for the numbers from 326 to 331 inclusive.

SECTION 15

A Second Look at a Problem from Romania

The following problem is discussed in my *More Mathematical Morsels* (pages 93–95); it appeared on the Third Selection Test given to candidates for the 1978 International Olympiad Team of Romania.

> Prove that the value of
>
> $$x\sqrt{2} + y\sqrt{3} + z\sqrt{5},$$
>
> where x, y, z are integers, not all zero, can be made arbitrarily close to zero.
>
> The following beautiful solution is due to Professor Liang-shin Hahn of the University of New Mexico.

It is easy to see that there are any number of triples (x, y, z) which yield a value of $x\sqrt{2} + y\sqrt{3} + z\sqrt{5}$ between 0 and $\sqrt{2}$. One way of generating such triples is to put $z = 0$, let y be a negative integer, thus making $y\sqrt{3} + z\sqrt{5}$ a negative quantity, and then determine the number x of $\sqrt{2}$'s that need to be added to first reach a positive value. For example, for $y = -3$, we obtain

$$-3\sqrt{3} = -3(1.732) \text{ approximately} = -5.196,$$

which requires four $\sqrt{2}$'s to be added to reach a positive value:

$$-5.196 + 4\sqrt{2} = -5.196 + 5.656 \text{ approximately} = .46,$$

a number between 0 and $\sqrt{2}$. Thus $(4, -3, 0)$ is such a triple.

Since repeatedly adding $\sqrt{2}$ advances the value of $y\sqrt{3}$ by steps of size $\sqrt{2}$, the only way the increasing total can avoid the open interval $(0, \sqrt{2})$ is for it to take the values at both endpoints, first 0 and then $\sqrt{2}$. But if

$$x\sqrt{2} + y\sqrt{3} = 0, \quad \text{for nonzero integers } x \text{ and } y,$$

then

$$x\sqrt{2} = -y\sqrt{3}$$

and

$$2x^2 = 3y^2,$$

which is impossible, for in their prime decompositions, the integer on the left has an odd number of factors equal to 2, while the integer on the right has an even number of them. Thus all these triples yield values of $x\sqrt{2}+y\sqrt{3}+z\sqrt{5}$ which lie in the **interior** of the interval.

Obviously, changes in y lead to different values of x, implying that there is an infinity of such triples $(x, y, 0)$. Moreover, no two of these triples yield the same value of $x\sqrt{2} + y\sqrt{3} + z\sqrt{5}$. Suppose to the contrary that some different triples $(x_1, y_1, 0)$, $(x_2, y_2, 0)$ give

$$x_1\sqrt{2} + y_1\sqrt{3} = x_2\sqrt{2} + y_2\sqrt{3};$$

then

$$(x_1 - x_2)\sqrt{2} = (y_2 - y_1)\sqrt{3},$$

and

$$(x_1 - x_2)^2 \cdot 2 = (y_2 - y_1)^2 \cdot 3.$$

Since the triples are different, not both $(x_1 - x_2)$ and $(y_1 - y_2)$ can be zero, making the equation impossible for the same reason $2x^2 = 3y^2$ is impossible.

Thus there is an infinity of different values of $x\sqrt{2} + y\sqrt{3} + z\sqrt{5}$ in the open interval $(0, \sqrt{2})$. Consequently these values must be packed together arbitrarily closely; that is to say, for any preassigned positive real number ε, however small, the magnitude of the difference between some pair of them is less than ε:

$$|(x_1\sqrt{2} + y_1\sqrt{3}) - (x_2\sqrt{2} + y_2\sqrt{3})| < \varepsilon.$$

Simplifying, this yields

$$|(x_1 - x_2)\sqrt{2} + (y_1 - y_2)\sqrt{3}| < \varepsilon,$$

implying that the triple $(x_1 - x_2, y_1 - y_2, 0)$, whose components are not all zero, gives a value of $x\sqrt{2} + y\sqrt{3} + z\sqrt{5}$ which is closer to zero than ε. The conclusion follows.

SECTION 16

32 Miscellaneous Problems

There are some pretty easy problems in this set; however, they are none the less attractive for being simple.

1. (From the 1990 Bulgarian Olympiad)

   The first 1990 positive integers are written in order in a row to form a long string of digits S_0:

$$S_0 = 1\,2\,3\,4\,5\,6\,7\,8\,9\,1\,0\,1\,1\,1\,2\,1\,3\,1\,4\,1\,5 \ldots 1\,9\,8\,9\,1\,9\,9\,0.$$

   Then the digits in all the even-numbered places are deleted and the gaps closed up to give the string

$$S_1 = 1\,3\,5\,7\,9\,0\,1\,2\,3\,4 \ldots 9\,9\,9\,0.$$

   Next the digits in the odd-numbered places in S_1 are deleted and the gaps closed to give

$$S_2 = 3\,7\,0\,2\,4 \ldots.$$

   This thinning by alternately deleting the digits in the even- numbered and odd-numbered places is continued until there is only one digit left. What is this last digit?

There are nine single-digit integers in S_0, 90 2-digit integers, 900 3-digit integers, and 991 4-digit integers, for a total of

$$9 + 180 + 2700 + 3964 = 6853 \text{ digits.}$$

Each round of deletions reduces the number of digits essentially in half, and since 6853 lies between 2^{12} and 2^{13}, 13 rounds of deletions will reduce the string to its final digit. Fortunately it is not too difficult to keep track of the **original positions** of the digits in the string S_k that is obtained **after** k rounds of deletions.

STRING	**ORIGINAL POSITIONS** OF THE DIGITS	
S_0	1 2 3 4 5 6 7 8 9 10 11 12 13 14 15...	
S_1	1 3 5 7 9 11 13 ...	(with increment of size 2)
S_2	3 7 11 15 ...	(with increment of size 2^2)
S_3	3 11 19 27 ...	(with increment of size 2^3)
S_4	11 27 43 59 ...	(with increment of size 2^4)
S_5	11 43 75 107 ...	(with increment of size 2^5)
S_6	43 107 171 235 ...	(with increment of size 2^6)

Since every second digit in S_k is deleted in producing S_{k+1}, the sequence of **original positions** of the digits in S_k increases from digit to digit by 2^k (since the first string is S_0).

Now, in going from S_{2k} to S_{2k+1}, it is the even-numbered places that are deleted, after which the deletion of the odd-numbered places gives S_{2k+2}: if

$$S_{2k} = a, b, c, d, e, f, g, \ldots,$$

then

$$S_{2k+1} = a, c, e, g, \ldots,$$

and

$$S_{2k+2} = c, g, \ldots$$

Therefore S_{2k} and S_{2k+1} begin with the same digit, and the first digit in S_{2k+2} is the **third** digit in S_{2k}. Hence

(the **original position** of the **first** digit in S_{2k+2})
= (the original position of the **third** digit in S_{2k})
= (the original position of the first digit in S_{2k})
$+ 2 \cdot$ (the increment for the sequence S_{2k}),
= (the original position of the first digit in S_{2k}) $+ 2 \cdot 2^{2k}$
= (the original position of the first digit in S_{2k}) $+ 2^{2(k+1)-1}$.

It follows, then, that if

(the original position of the first digit in S_{2k})
$= 1 + 2 + 2^3 + 2^5 + \cdots + 2^{2k-1}$,

then

$$\text{(the original position of the first digit in } S_{2k+2})$$
$$= (1 + 2 + 2^3 + \cdots + 2^{2k-1}) + 2^{2(k+1)-1},$$

and since the series $1 + 2 + 2^3 + 2^5 + \cdots + 2^{2k-1}$ gives the correct value for $k = 1$, it holds for all $k \geq 1$ by induction. Hence, for $k \geq 1$,

$$\text{the original position of the first digit in } S_{2k} = 1 + 2 + 2^3 + 2^5 + \cdots + 2^{2k-1}.$$

Now, S_{11} begins with the same digit as S_{10}, namely the digit whose original position is

$$1 + 2 + 2^3 + 2^5 + 2^7 + 2^9 = 1 + 2 + 8 + 32 + 128 + 512 = 683,$$

and the increment in its sequence of original positions is $2^{11} = 2048$. Hence the original places of the digits of S_{11} are

$$S_{11} : 683, 2731, 4779, 6827.$$

At this point it is the odd-numbered places that are deleted, and so for S_{12}, the original positions are

$$S_{12} : 2731, 6827,$$

and finally, after 13 rounds, S_{13} holds just the 2731st digit of S_0.

S_0 begins with nine 1-digit numbers and 90 2-digit numbers, accounting for the first 189 places. Thus we need to count $2731 - 189 = 2542$ places along the 3-digit integers. Since $\frac{2542}{3} = 847\frac{1}{3}$, the required digit is the **first** digit of the 848th integer, which is $99 + 848 = 947$. The required digit, then, is 9.

2. (From proposals submitted for the 1985 Canadian Olympiad)

For each positive integer n, prove there are two **consecutive** positive integers each of which is the product of n positive integers > 1.

Observe that the product of the n powers of 2,

$$2 \cdot 2^2 \cdot 2^{2^2} \cdot 2^{2^3} \cdot \cdots \cdot 2^{2^{n-2}} \cdot 2^{2^{n-1}+1}$$
$$= 2^{(1+2+2^2+2^3+\cdots+2^{n-1})+1} = 2^{(2^n-1)+1} = 2^{2^n}.$$

This suggests that we see whether the integer $2^{2^n} - 1$ is the product of n factors > 1.

Repeatedly factoring as the difference of two squares, we get

$$\begin{aligned}2^{2^n} - 1 &= (2^{2^{n-1}} + 1)(2^{2^{n-1}} - 1)\\ &= (2^{2^{n-1}} + 1)(2^{2^{n-2}} + 1)(2^{2^{n-2}} - 1)\\ &\quad\cdots\cdots\cdots\cdots\cdots\cdots\cdots\cdots\\ &= (2^{2^{n-1}} + 1)(2^{2^{n-2}} + 1)\ldots(2^{2^0} + 1)(2^{2^0} - 1),\end{aligned}$$

when the last factor, the $(n+1)$st factor, is simply $2-1=1$. Thus $2^{2^n}-1$ does indeed factor into n positive integers > 1, making $2^{2^n}-1$ and 2^{2^n} a satisfactory pair for each value of n.

3. (From proposals for a Freshman Contest at the University of Waterloo)

Determine all integral solutions of $a^2 + b^2 + c^2 = a^2b^2$.

Clearly $(0, 0, 0)$ is a solution. Suppose (a, b, c) is a solution in which $a > 0$. Since a square is congruent (mod 4) to 0 or 1,

$$a^2b^2 \equiv 0 \text{ or } 1 \pmod 4.$$

Thus

$$a^2 + b^2 + c^2 \equiv 0 \text{ or } 1 \pmod 4,$$

and so not more than one of a^2, b^2, c^2 can be $\equiv 1 \pmod 4$, in particular not both a^2 and b^2, implying that a^2b^2 must be $\equiv 0 \pmod 4$. In this case,

$$a^2 + b^2 + c^2 = a^2b^2 \equiv 0 \pmod 4,$$

and we have

$$a^2 \equiv b^2 \equiv c^2 \equiv 0 \pmod 4;$$

that is to say, each of a, b, c, is even. Suppose $a = 2a_1$, $b = 2b_1$, $c = 2c_1$. Then $a_1 > 0$, and

$$4a_1^2 + 4b_1^2 + 4c_1^2 = 16a_1^2b_1^2,$$

and

$$a_1^2 + b_1^2 + c_1^2 = 4a_1^2b_1^2.$$

Thus if (a, b, c) is a solution of $a^2 + b^2 + c^2 = a^2b^2$ in which $a > 0$, then (a_1, b_1, c_1) is a solution of $a^2 + b^2 + c^2 = 4a^2b^2$ in which $a_1 > 0$. Then, as above, we have

$$a_1^2 \equiv b_1^2 \equiv c_1^2 \equiv 0 \pmod 4,$$

and letting $a_1 = 2a_2, b_1 = 2b_2, c_1 = 2c_2$, we see that (a_2, b_2, c_2) is a solution of

$$a^2 + b^2 + c^2 = 4^2 a^2 b^2, \quad \text{where } a_1 > a_2 > 0.$$

Continuing to apply this argument establishes the existence of an endless sequence of decreasing positive integers $\{a > a_1 > a_2 > \cdots > a_n > \cdots\}$, each one-half the previous integer, arising from solutions (a_n, b_n, c_n) of the equations

$$a^2 + b^2 + c^2 = 4^n a^2 b^2, \quad n = 0, 1, 2, \ldots.$$

But, starting from any positive integer a, an **infinite descent** within the positive integers is impossible, and hence there can be no solution of $a^2 + b^2 + c^2 = a^2 b^2$ with $a > 0$. Now, if either (a, b, c) or $(-a, b, c)$ is a solution, then the other is, too. Thus no solution with $a > 0$ also means no solution with $a < 0$, implying $a = 0$ in all solutions. Similarly b must be 0, and therefore every solution is of the form $(0, 0, c)$ and must satisfy

$$a^2 + b^2 + c^2 = 0^2 + 0^2 + c^2 = 0,$$

implying that $(0, 0, 0)$ is the only solution.

4. (From the 1989 Competition of the Republic of Yugoslavia)

   Express

$$f(x) = (x^2 + 1)(x^2 + 4)(x^2 - 2x + 2)(x^2 + 2x + 2)$$

   as the sum of the squares of two polynomials having integral coefficients.

Since

$$\begin{aligned}(x^2 - 2x + 2)(x^2 + 2x + 2) &= [(x^2 + 2) + 2x][(x^2 + 2) - 2x] \\ &= (x^2 + 2)^2 - 4x^2 \\ &= x^4 + 4,\end{aligned}$$

we have

$$f(x) = (x^2 + 1)(x^2 + 4)(x^4 + 4).$$

Now,

$$\begin{aligned}(a^2 + b^2)(c^2 + d^2) &= a^2c^2 + a^2d^2 + b^2c^2 + b^2d^2 \\ &= (a^2c^2 + 2abcd + b^2d^2) + (a^2d^2 - 2abcd + b^2c^2) \\ &= (ac + bd)^2 + (ad - bc)^2.\end{aligned}$$

Therefore

$$\begin{aligned}(x^2+1)(x^2+4) &= (x^2+2)^2+(2x-x)^2\\ &= (x^2+2)^2+x^2,\end{aligned}$$

and finally,

$$\begin{aligned}f(x) &= [(x^2+2)^2+x^2](x^4+4)\\ &= [(x^2+2)x^2+2x]^2+[2(x^2+2)-x^3]^2\\ &= (x^4+2x^2+2x)^2+(-x^3+2x^2+4)^2\end{aligned}$$

or

$$f(x) = (x^4+2x^2+2x)^2+(x^3-2x^2-4)^2.$$

5. (From the 1990 American Invitational Mathematics Examination)

   The increasing sequence $S = \{2, 3, 5, 6, 7, 10, 11, \ldots\}$ consists of all positive integers which are neither a perfect square nor a perfect cube. What is the 500th term of S?

Clearly the required term n is unique and greater than 500. Since the squares and cubes thin out as they increase, n is not likely to be very much greater than 500. In fact, the number of squares less than 500 is 22 ($22^2 = 484$) and the number of cubes is only 7 ($8^3 = 512$). Thus, up to 500 not more than 29 positive integers have failed to qualify for membership in S. Actually, since the sixth powers 1 and 64 are both squares and cubes, only 27 positive integers < 500 are not in S, making 500 itself the 473rd term in S. Advancing 27 terms from 500 would take us to 528, since the cube 512 is missing, and since 512 is the only positive integer from 500 to 528 which is not in S, the 500th term of S is indeed 528. (This cuts things pretty close, for the next integer not in S is $529 = 23^2$.)

6. (From the 1988 Manitoba Mathematical Contest)

   ΔABC is isosceles with $AB = AC$. AB is extended an arbitrary distance to D and DF is drawn to AC so that its midpoint M lies on BC (Figure 1). Prove $BD = CF$.

Let $DB = x$ and $CF = y$. Now, extending AB has unbalanced the figure and so let us restore a balance by extending AC the same distance x to N

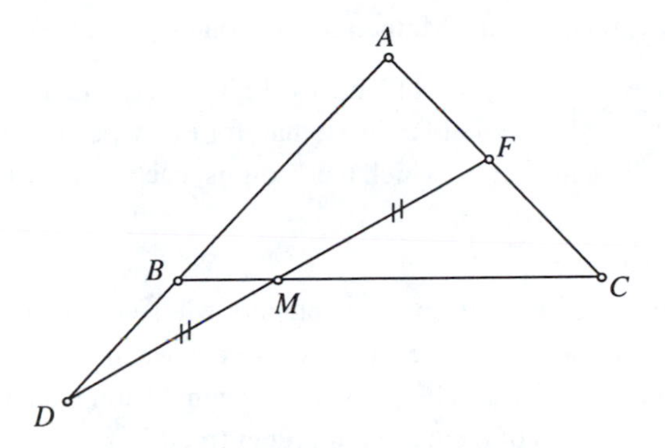

FIGURE 1

(Figure 2). Then triangles ABC and ADN are isosceles and have the same base angles, making DN parallel to BC, and we have

$$\triangle DMN = \triangle DCN \text{ (on the same base and between the same parallels).}$$

But because NM is a median of $\triangle DFN$,

$$\triangle DMN = \frac{1}{2}\triangle DFN.$$

Hence

$$\triangle DCN = \frac{1}{2}\triangle DFN,$$

implying DC is also a median, making $x = y$, as desired.

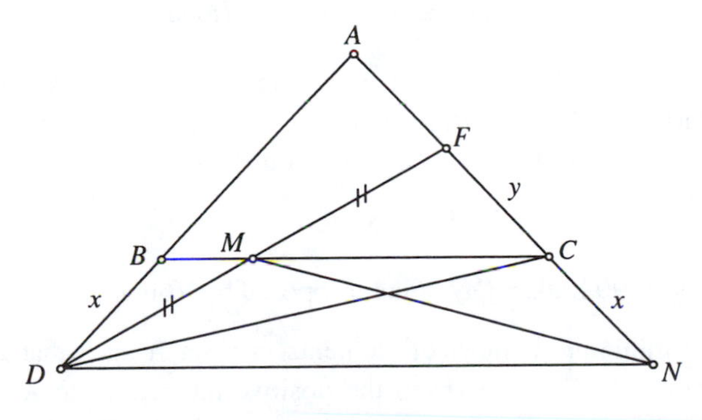

FIGURE 2

7. (From the 1990 Manitoba Mathematical Contest)

   The positive integer $a_3 = 112$ has 3 digits, each either 1 or 2, and it is divisible by 2^3 ($112 = 8 \cdot 14$). Prove that, for every positive integer n, there exists a positive integer a_n which has n digits, each either 1 or 2, and which is divisible by 2^n.

Consider the 2^n n-digit positive integers whose digits are each either 1 or 2. If no two of them yield the same remainder when divided by 2^n, one of the remainders would have to be zero and the conclusion would follow. Let us show, then, that no two of them are congruent (mod 2^n).

Proceeding indirectly, suppose some two different such n-digit integers **are** congruent (mod 2^n):

$$a_n a_{n-1} \dots a_1 \equiv b_n b_{n-1} \dots b_1 \pmod{2^n}.$$

Then their last digits would have to be the same in order to make their difference divisible by the even number 2^n: $a_1 = b_1$. Thus (mod 2^n)

$$10a_n a_{n-1} \dots a_2 + a_1 \equiv 10b_n b_{n-1} \dots b_2 + b_1$$

and

$$10a_n a_{n-1} \dots a_2 \equiv 10b_n b_{n-1} \dots b_2,$$

from which, dividing through by 2, we obtain

$$5a_n a_{n-1} \dots a_2 \equiv 5b_n b_{n-1} \dots b_2 \pmod{2^{n-1}}.$$

Since 2 and 5 are relatively prime, this yields

$$a_n a_{n-1} \dots a_2 \equiv b_n b_{n-1} \dots b_2 \pmod{2^{n-1}},$$

which in turn implies $a_2 = b_2$ by the same argument which gave us $a_1 = b_1$.

Similarly, $a_3 = b_3$, and in general $a_k = b_k$, yielding the contradiction that $a_n a_{n-1} \dots a_1$ and $b_n b_{n-1} \dots b_1$ are not different after all, completing the solution.

8. (From the 1990 Balkan Olympiad; proposed by Romania)

   Find the minimum number of elements in a set A such that there exists a function $f : N \to A$ (from the positive integers N to A) having the property that $f(i) \neq f(j)$ whenever $|i - j|$ is a prime number.

If $f(n)$ is given by the remainder r obtained by dividing n by 4, that is, $f(n) = r$, where $n \equiv r \pmod 4$, then $f(i)$ and $f(j)$ can be equal only when $|i - j|$ is a multiple of 4, implying $f(i)$ and $f(j)$ are unequal whenever $|i - j|$ is a prime number. Thus four elements suffice for the set A.

On the other hand, since each two of the integers 1, 3, 6, and 8 differ by a prime number, no two of $f(1)$, $f(3)$, $f(6)$, and $f(8)$ can be the same, implying A must have at least four members.

Thus the minimum number of elements in A is four.

9. (From the 1989–1990 Iranian High School Mathematical Competition)

The sequence $a_1, a_2, \ldots$ is defined by $a_1 = 1$, $a_2 = 2$, and for $n \geq 2$,

$$a_{n+1} = 1 + a_1 a_2 \ldots a_{n-1} + (a_1 a_2 \ldots a_{n-1})^2.$$

Prove that, as $n \to \infty$,

$$\lim \left(\frac{1}{a_1} + \frac{1}{a_2} + \cdots + \frac{1}{a_n} \right) = 2.$$

With the sequence defined by such an unwieldy condition, it seems advisable to see whether there isn't a simpler definition. The sequence begins

$$1, 2, 3, 7, 43, 1807, \ldots,$$

suggesting that each term is $1 +$ (the product of all the preceding terms). From the given definition we have

$$a_{n+1} = 1 + a_1 a_2 \ldots a_{n-1}(1 + a_1 a_2 \ldots a_{n-1}),$$

and indeed, if the bracket $(1 + a_1 a_2 \ldots a_{n-1})$ is equal to a_n, then a_{n+1} would have the same form:

$$a_{n+1} = 1 + a_1 a_2 \ldots a_{n-1}(a_n).$$

Since this is clearly valid for the first few terms, the relation holds by induction for the whole sequence.

Next, consider the partial sums

$$S_n = \frac{1}{a_1} + \frac{1}{a_2} + \cdots + \frac{1}{a_n}.$$

The first four values are

$$S_1 = 1, \quad S_2 = \frac{3}{2}, \quad S_3 = \frac{11}{6}, \quad S_4 = \frac{83}{42},$$

suggesting the formula

$$S_n = 2 - \frac{1}{a_{n+1} - 1},$$

which is also easily proved by induction as follows.

First, observe that from our improved definition

$$a_n = 1 + a_1 a_2 \ldots a_{n-1}$$

we get

$$a_1 a_2 \ldots a_{n-1} = a_n - 1,$$

and so

$$\begin{aligned} a_{n+1} &= 1 + a_1 a_2 \ldots a_{n-1} a_n \\ &= 1 + (a_n - 1) a_n, \end{aligned}$$

giving

$$a_{n+1} - 1 = (a_n - 1) a_n.$$

Now, if

$$S_{n-1} = 2 - \frac{1}{a_n - 1},$$

then

$$\begin{aligned} S_n &= S_{n-1} + \frac{1}{a_n} = \left[2 - \frac{1}{a_n - 1}\right] + \frac{1}{a_n} = 2 - \left[\frac{a_n - (a_n - 1)}{(a_n - 1) a_n}\right] \\ &= 2 - \frac{1}{(a_n - 1) a_n} = 2 - \frac{1}{a_{n+1} - 1}, \quad \text{as desired.} \end{aligned}$$

Since a_n clearly grows beyond all bounds, it follows that, as $n \to \infty$,

$$S = \lim S_n = 2.$$

10. (From the 1983 Putnam Competition)

The hands of an accurate clock have lengths 3 and 4. Find the distance between the tips of the hands when that distance is increasing most rapidly.

The tips A and B of the hands and the center O of the face form a triangle AOB having sides $OA = 3$, $OB = 4$, a side of variable length $AB = x$, and a variable $\angle AOB = \theta$ (Figure 3).

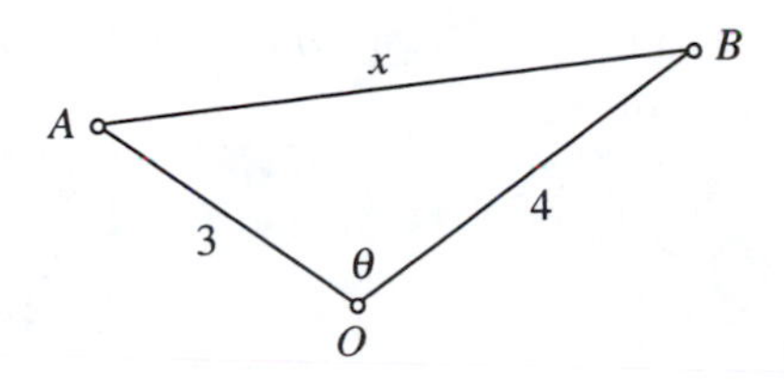

FIGURE 3

By the law of cosines, we have

$$x = (3^2 + 4^2 - 2 \cdot 3 \cdot 4 \cdot \cos\theta)^{1/2} = (25 - 24\cos\theta)^{1/2};$$

hence

$$\frac{dx}{d\theta} = \frac{24\sin\theta}{2(25 - 24\cos\theta)^{1/2}} = \frac{12\sin\theta}{(25 - 24\cos\theta)^{1/2}}.$$

Now, for a maximum value of $\frac{dx}{d\theta}$, the second derivative must vanish, in which case

$$\frac{d^2x}{d\theta^2} = 12\left[\frac{\cos\theta}{(25 - 24\cos\theta)^{1/2}} - \frac{\sin\theta \cdot (24\sin\theta)}{2(25 - 24\cos\theta)^{3/2}}\right] = 0,$$

$$\frac{\cos\theta}{(25 - 24\cos\theta)^{1/2}} - \frac{12\sin^2\theta}{(25 - 24\cos\theta)^{3/2}} = 0,$$

$$(\cos\theta)(25 - 24\cos\theta) = 12\sin^2\theta,$$

$$25\cos\theta - 24\cos^2\theta = 12\sin^2\theta,$$

$$12\cos^2\theta - 25\cos\theta + 12 = 0,$$

$$(3\cos\theta - 4)(4\cos\theta - 3) = 0,$$

and

$$\cos\theta = \frac{4}{3}, \frac{3}{4}.$$

Thus $\cos\theta$ would have to be $\frac{3}{4}$, making $x = (25 - 24\cos\theta)^{1/2} = \sqrt{7}$.

In order to see that this gives a maximum value of $\frac{dx}{d\theta}$, observe that, for $\cos\theta = \frac{3}{4}$, we have $\sin\theta = \frac{\sqrt{7}}{4} = .661\ldots$ (Figure 4), giving $\frac{dx}{d\theta} = \frac{3\sqrt{7}}{\sqrt{7}} = 3$, and for a slightly smaller θ, say $\theta = \sin^{-1}\left(\frac{3}{5}\right)$, we have

$$\frac{dx}{d\theta} = \frac{12\left(\frac{3}{5}\right)}{\left(25 - 24 \cdot \frac{4}{5}\right)^{1/2}} = \frac{36}{\sqrt{145}} < \frac{36}{\sqrt{144}} = 3.$$

Thus the desired distance is $x = \sqrt{7}$, at which point $\angle OAB$ is a right angle (Figure 5).

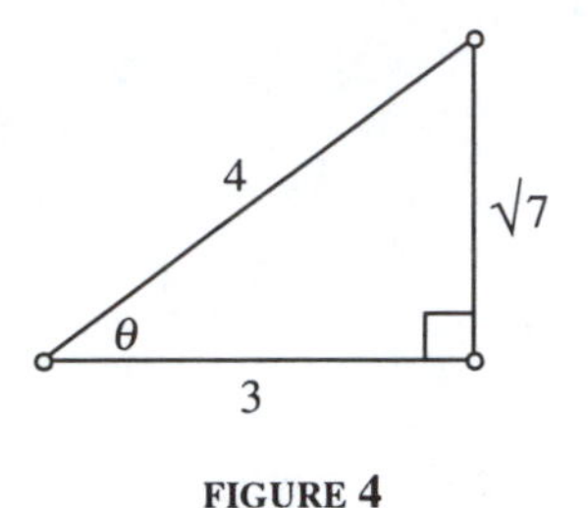

FIGURE 4

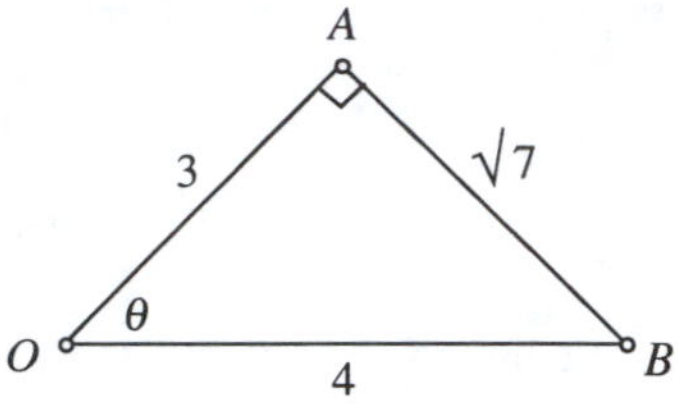

FIGURE 5

11. (From the Mathematics Contest of the Puig Adam Society of Teachers for ages 14 and 15 (Spain))

    Find the maximum value of the positive integer n for which n^2+2 divides $f(n) = (n+1)(n^4+2n) + 3(n^3+57)$.

Multiplying out, we get

$$f(n) = (n+1)(n^4+2n) + 3(n^3+57) = n^5 + n^4 + 3n^3 + 2n^2 + 2n + 171.$$

Although we are not concerned with n^2+2 being a literal factor of $f(n)$, if we divide $f(n)$ by n^2+2 we see that

$$f(n) = (n^2+2)(n^3+n^2+n) + 171.$$

Consequently,

$$n^2+2 \mid f(n) \quad \text{implies} \quad n^2+2 \mid 171,$$

for which the maximum n is obviously 13.

12. (From the 1975 Special-K Contest at the University of Waterloo)

    The product of all the positive integral divisors of the positive integer n is $P = 2^{120} \cdot 3^{60} \cdot 5^{90}$. Find n.

Clearly n must be of the form $2^a \cdot 3^b \cdot 5^c$ and the divisors of n are

$$d = 2^x \cdot 3^y \cdot 5^z, \quad \text{where} \quad 0 \le x \le a,\ 0 \le y \le b,\ 0 \le z \le c.$$

Now, each x appears in a divisor d with each of the $(b+1)(c+1)$ choices for the pair y and z; hence the total exponent of 2 in the product P is

$$(b+1)(c+1)(0+1+2+\cdots+a) = 120,$$

giving

$$a(a+1)(b+1)(c+1) = 240.$$

Similarly we have

$$b(a+1)(b+1)(c+1) = 120$$

and

$$c(a+1)(b+1)(c+1) = 180.$$

Dividing these equations we get

$$b = \frac{1}{2}a \quad \text{and} \quad c = \frac{3}{4}a,$$

and so

$$a(a+1)\left(\frac{1}{2}a+1\right)\left(\frac{3}{4}a+1\right) = 240.$$

Since $c = \frac{3}{4}a$, a must be divisible by 4, suggesting a might be 4. Checking, we find that $a = 4$ does satisfy the equation

$$4(5)(3)(4) = 240.$$

Hence $a = 4$, giving $b = 2$ and $c = 3$, making

$$n = 2^4 \cdot 3^2 \cdot 5^3 = 18000.$$

13. (From the proposals for the 1984 Canadian Olympiad)

    Let n be any positive integer. Prove that the decimal representation of some multiple of n begins with the digits 1984.

If n has k digits, let $m = 1984 \cdot 10^k = 198400\ldots00$. Then the n integers

$$m, m+1, m+2, \ldots, m+n-1$$

all begin with 1984, and since every string of n consecutive integers contains a multiple of n, the conclusion follows.

14. (From the proposals for the 1984 Canadian Olympiad)

    For every positive integer $n > 1$ prove there exist n factorials, each > 1, whose product is also a factorial: $x_1!\,x_2!\ldots x_n! = y!$.

Since $(k!)! = (k!)(k!-1)!$, it is easy to see that the claim is valid for $n+1$ whenever it is valid for n: if

$$x_1!\,x_2!\ldots x_{n-1}! = y!,$$

then

$$x_1!\,x_2!\ldots x_{n-1}!\,(y!-1)! = (y!)(y!-1)! = (y!)!.$$

The trouble is to find a particular solution of $x!y! = z!$ with which to seed the induction. However, since $8\cdot 9\cdot 10 = 720 = 6!$, we have

$$10! = 7!(8\cdot 9\cdot 10) = 7!\cdot 6!,$$

and the desired result follows by induction.

15. (From the 1989–1990 U.S.A. Mathematical Talent Search)

    For positive integers m and n, prove that

$$3^m + 3^n + 1 \text{ is never a perfect square.}$$

Clearly $3^m + 3^n + 1$ is odd. Suppose, for some positive integer k, that

$$3^m + 3^n + 1 = (2k+1)^2.$$

Then

$$3^m + 3^n = 4k^2 + 4k = 4k(k+1),$$

and since one of the consecutive integers k, $k+1$ must be even, it follows that $3^m + 3^n$ is divisible by 8. But

$$3^2 \equiv 1 (\text{mod } 8),$$

implying

$$3^{2t} \equiv 1 \text{ and } 3^{2t+1} \equiv 3 (\text{mod } 8) \text{ for all } t.$$

Hence

$$3^m + 3^n \equiv 1+1,\ 1+3, \quad \text{or} \quad 3+3 (\text{mod } 8),\ \textbf{never } 0,$$

and the conclusion follows by contradiction.

16. (From the 1989–1990 U.S.A. Mathematical Talent Search)

If n is a positive integer > 11, prove that

$$n^2 - 19n + 89 \text{ is never a perfect square.}$$

(a) Suppose

$$n^2 - 19n + 89 = k^2.$$

Then

$$n^2 - 19n + (89 - k^2) = 0$$

and

$$n = \frac{19 \pm \sqrt{361 - 356 + 4k^2}}{2} = \frac{19 \pm \sqrt{4k^2 + 5}}{2}.$$

Since n is an integer, $4k^2 + 5$ must be some perfect square $(2t + 1)^2$, giving

$$4k^2 + 5 = 4t^2 + 4t + 1,$$

and

$$k^2 + 1 = t^2 + t.$$

Therefore

$$k^2 + k \geq k^2 + 1 = t^2 + t,$$

implying $k \geq t$.

Suppose $k > t$, that is, $k = t + s$ for some integer $s \geq 1$. Then

$$k^2 + 1 = t^2 + 2ts + s^2 + 1 = t^2 + t,$$

giving

$$2ts + s^2 + 1 = t.$$

But

$$2ts + s^2 + 1 > 2ts \geq 2t > t,$$

showing this equation is impossible.

Hence $k = t$, and we have

$$k^2 + 1 = t^2 + t = k^2 + k, \text{ implying } k = 1.$$

Thus

$$4k^2 + 5 = 9, \quad \text{and} \quad n = \frac{19 \pm 3}{2} = 11 \text{ or } 8.$$

For $n > 11$, then,

$$n^2 - 19n + 89 \text{ is never a perfect square.}$$

This is a perfectly acceptable solution, but it doesn't compare with the following beautiful solution of George Berzsenyi (Rose-Hulman Institute of Technology, Terre Haute, Indiana), which is simplicity itself.

(b) Since the expansion of $(n-9)^2$ contains n^2-18n and that of $(n-10)^2$ contains n^2-20n, thus bracketing n^2-19n, let us compare the given function with each of these squares.

(i) $(n-9)^2 - (n^2 - 19n + 89) = n - 8 > 0$ for $n > 11$;
(ii) $(n^2 - 19n + 89) - (n - 10)^2 = n - 11 > 0$ for $n > 11$.

Thus, for $n > 11$, $n^2 - 19n + 89$ always lies **between consecutive squares** and is therefore never one itself.

17. (From the 1989 American Invitational Mathematics Examination)

Ten points are marked around a circle. How many different convex polygons of three or more sides are there having vertices among the ten points?

At first glance one might think the requirement that the polygons be convex is a complicating restriction. In reality, the condition is our salvation, for the problem is considerably more involved without it:

> once the vertices have been selected, a convex polygon can be determined in just one way—by taking the vertices in cyclic order around the circle; otherwise some side would split the vertices and force the polygon to cross itself in order to pick up all the chosen vertices.

Thus the number of convex n-gons is simply $\binom{10}{n}$ and the total number of polygons is

$$\begin{aligned}
&\binom{10}{3}+\binom{10}{4}+\cdots+\binom{10}{10} \\
&\quad=\left[\binom{10}{0}+\binom{10}{1}+\binom{10}{2}+\cdots+\binom{10}{10}\right]-\binom{10}{0}-\binom{10}{1}-\binom{10}{2} \\
&\quad=2^{10}-1-10-45 \\
&\quad=1024-56 \\
&\quad=968.
\end{aligned}$$

18. (From the 1990 Manitoba Mathematical Contest)

    If two of the altitudes of a triangle have lengths 6 and 12, prove that the third altitude must exceed 4.

Let the triangle be ABC with sides of lengths a, b, c (Figure 6), and let the length of the third altitude be x. Now, the area of $\triangle ABC$ is

$$\frac{1}{2}\cdot 6a=\frac{1}{2}\cdot 12c, \quad \text{implying} \quad a=2c.$$

Similarly,

$$\frac{1}{2}\cdot bx=6c, \quad \text{giving} \quad x=\frac{12c}{b}.$$

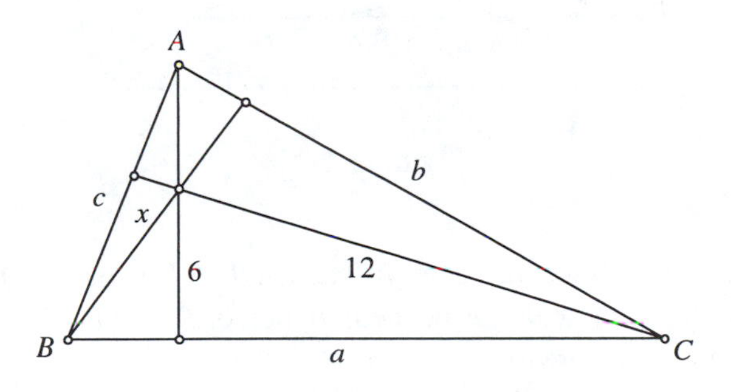

FIGURE 6

Now, the law of cosines yields

$$\begin{aligned}
b^2&=a^2+c^2-2ac\cos B \\
&=4c^2+c^2-4c^2\cos B \\
&<9c^2, \quad \text{since } \cos B>-1.
\end{aligned}$$

Hence

$$\frac{c}{b} > \frac{1}{3}$$

and thus

$$x = \frac{12c}{b} > 4.$$

19. (From the 1990 Manitoba Mathematical Contest)

    BD and CE are medians in $\triangle ABC$ and P and Q are the midpoints of BE and CD, respectively (Figure 7). Prove the engaging property that PQ is trisected by BD and CE, that is,

$$PR = RS = SQ.$$

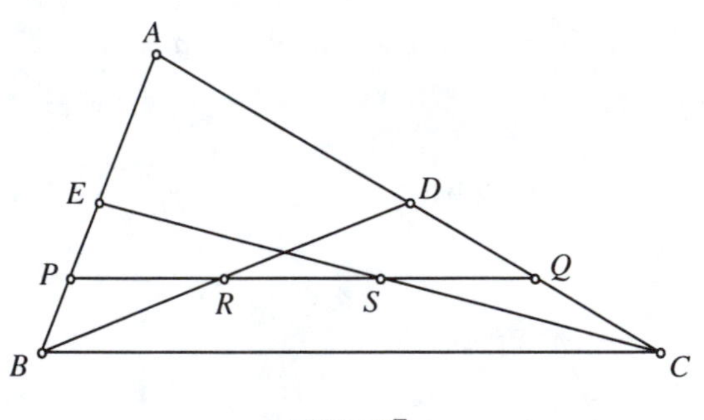

FIGURE 7

Since D and E bisect the sides AC and AB, DE is parallel to BC and half as long. Therefore, letting F be the midpoint of BC, $EBFD$ and $EFCD$ are parallelograms (Figure 8).

Clearly P and Q divide the sides AB and AC in the ratio of 3:1 and hence PQ is also parallel to BC. Moreover, since P bisects BE, PQ runs along halfway between BC and DE and bisects every transversal from DE to BC. Thus the point of intersection R of PQ and diagonal BD is the midpoint of BD, making R the center of parallelogram $EBFD$. Similarly, S is the center of $EFCD$, and since any transversal across a parallelogram through its center is itself bisected there, we have the desired $PR = RS = SQ$.

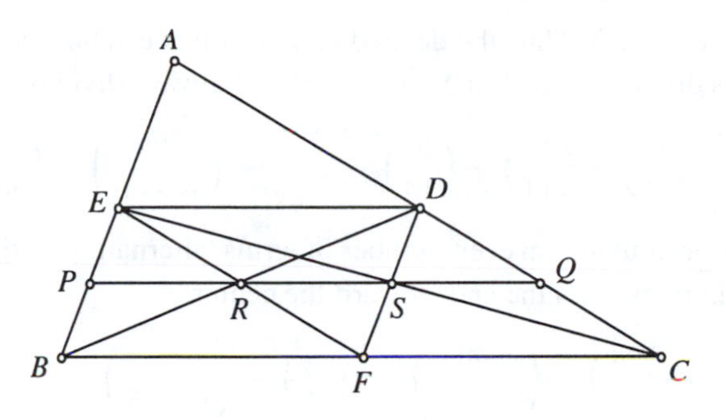

FIGURE 8

20. (From the 1983 Putnam Contest)

Let k be a positive integer and let $m = 6k - 1$. Also, let

$$S(m) = \sum_{j=1}^{2k-1} (-1)^{j+1} \binom{m}{3j-1}$$
$$= \binom{m}{2} - \binom{m}{5} + \binom{m}{8} - \binom{m}{11} + - \cdots + \binom{m}{m-3}.$$

For example, for $k = 3$, we get

$$S(17) = \binom{17}{2} - \binom{17}{5} + \binom{17}{8} - \binom{17}{11} + \binom{17}{14}.$$

Prove that $S(m)$ is never zero.

Observing a few cases, we have

$$K = 1: \quad S(5) = \binom{5}{2} = 10;$$
$$K = 2: \quad S(11) = \binom{11}{2} - \binom{11}{5} + \binom{11}{8} = 55 - 462 + 1165 = -242;$$
$$K = 3: \quad S(17) = \binom{17}{2} - \binom{17}{5} + \binom{17}{8} - \binom{17}{11} + \binom{17}{14}$$
$$= 136 - 6188 + 24310 - 12376 + 680 = 6562.$$

If a fixed positive integer n never divides $S(m)$, then $S(m)$ could never be zero. It appears that $n = 3$ is a possible candidate, so let's try to show that $S(m)$ is never divisible by 3.

We observe that, because $m = 6k - 1$,

$$\binom{m}{2} = \frac{(6k-1)(6k-2)}{2} = (6k-1)(3k-1)$$

is never divisible by 3. Thus the desired conclusion is established by showing that 3 always divides the rest of $S(m)$, i.e., that 3 always divides

$$-\binom{m}{5}+\binom{m}{8}-\binom{m}{11}+\binom{m}{14}-+\cdots-\binom{m}{m-6}+\binom{m}{m-3}.$$

This expression contains an even number of terms, alternating in sign, and they go together in pairs from the ends toward the center,

$$-\binom{m}{5}+\binom{m}{m-3}, \quad \binom{m}{8}-\binom{m}{m-6} \ldots,$$

i.e., pairs of the form

$$\pm\left[\binom{m}{3t-1}-\binom{m}{m-3t+3}\right], \quad t=2,3,4,\ldots.$$

Thus, if we can show that each pair is divisible by 3, our proof would be complete. We would like to show, then, that

$$N=\binom{m}{3t-1}-\binom{m}{m-3t+3}\equiv 0 \pmod 3, \quad t=2,3,4,\ldots.$$

We have

$$\begin{aligned} N &= \frac{m!}{(3t-1)!(m-3t+1)!}-\frac{m!}{(m-3t+3)!(3t-3)!} \\ &= \frac{m![(m-3t+2)(m-3t+3)-(3t-1)(3t-2)]}{(3t-1)!(m-3t+3)!}. \end{aligned}$$

Since $m=6k-1$, we have

$$m-3t+2\equiv m+2\equiv 1 \pmod 3$$

and

$$m-3t+3\equiv 2 \pmod 3,$$

implying that

$$\begin{aligned} &(m-3t+2)(m-3t+3)-(3t-2)(3t-1) \\ &\equiv 1\cdot 2-(-2)(-1)=0 \pmod 3, \end{aligned}$$

showing that the numerator is given by $m!(3T)$ for some integer T.

Finally, we shall show that all the factors 3 in the denominator divide into the factor $m!$ in the numerator, so that, when all the simplification is done, N still contains the factor 3 in $(3T)$. Our argument is based on the fact that the product of r consecutive integers is divisible by $r!$.

We have

$$\begin{aligned}
&\frac{m!}{(3t-1)!(m-3t+3)!}\\
&\quad=\frac{(6k-1)!}{(3t-1)!(6k-3t+2)!}\\
&\quad=\frac{(6k-1)(6k-2)\cdots(6k-3t+3)}{(3t-1)!}\\
&\quad=\frac{(6k-1)(6k-2)\cdots[6k-(3t-3)]}{(3t-1)(3t-2)(3t-3)!}\\
&\quad=\frac{1}{(3t-1)(3t-2)}\left[\frac{(6k-1)(6k-2)\cdots[6k-(3t-3)]}{(3t-3)!}\right],
\end{aligned}$$

(where the numerator of the second large factor is the product of $3t-3$ consecutive integers and therefore this factor reduces to an integer after division by $(3t-3)!$)

$$=\frac{A}{(3t-1)(3t-2)} \quad \text{for some integer } A.$$

Hence

$$N=\frac{A(3T)}{(3t-1)(3t-2)},$$

and since there is no factor 3 in the denominator, the simplification does not affect the factor 3 in the numerator, and we have $3 \mid N$, completing the argument.

21. Suppose you are presented with six balls, two of which are red, two are white, and two are blue. One ball of each color is light and the other is heavy. The light balls all have the same weight, and the heavy balls also weigh the same as each other. How can one determine which balls are light and which are heavy with just **two** weighings with an equal-arm balance?

The following neat solution is due to my colleague Ian McGee.

Let the balls be denoted by $r_1, r_2, w_1, w_2, b_1, b_2$, for red, white, and blue, and let the first weighing be

$$r_1+w_1 \quad \text{against} \quad r_2+b_1.$$

Either a balance is obtained or not. The two cases of imbalance are essentially the same and so we need consider only the cases (i) a balance and (ii) say the left pan (r_1+w_1) goes down.

(i) A Balance In this case, the weights of the balls on the left must be duplicated by the balls on the right—two heavy balls on the left imply two heavy balls on the right, etc. Since we know r_1 on the left is **not** the same as r_2 on the right, it must be that r_1 has the same weight as b_1 on the right, making w_1 the same as r_2, two facts that we may express by the equations

$$r_1 = b_1 \quad \text{and} \quad r_2 = w_1.$$

But we also know that w_1 and w_2 are different, and since $w_1 = r_2$, it follows that w_2 must be the same as r_1. Thus

$$r_1 = b_1 = w_2,$$

and similarly

$$r_2 = w_1 = b_2.$$

Therefore, weighing the two balls of any color against each other settles the entire affair.

(ii) An Imbalance ($r_1 + w_1$ outweighs $r_2 + b_1$) Let us denote light balls by L and heavy ones by H, e.g., $r_1 = H$ means r_1 is heavy. Now, if $r_2 = H$, then $r_2 + b_1$ would weigh at least $H + L$, which the left side could not overbalance because r_1 would be L (in view of $r_2 = H$). Thus it must be that $r_1 = H$ and $r_2 = L$. With the red balls sorted out, there are only four possible cases for the white and blue balls:

	w_1	w_2	b_1	b_2
(1)	H	L	H	L
(2)	H	L	L	H
(3)	L	H	H	L
(4)	L	H	L	H.

But we can get still more out of the first weighing. If $b_1 = H$, making the right side $r_2 + b_1 = L + H$, the left side could only overbalance it by weighing $H + H$, which would require $w_1 = H$. That is to say, we cannot have both $b_1 = H$ and $w_1 = L$, and so case (3) above is out of the running.

It remains to set up a second weighing whose three outcomes distinguish between possibilities (1), (2), and (4). Since the red balls are already settled, we can accomplish our goal by concentrating on the white and blue balls, setting off

$$w_1 + b_1 \quad \text{against} \quad w_2 + b_2.$$

By so doing, this pits

$$\begin{aligned} \text{in case (1):} &\quad H + H \text{ against } L + L, \\ \text{in case (2):} &\quad H + L \text{ against } L + H, \\ \text{and in case (4):} &\quad L + L \text{ against } H + H. \end{aligned}$$

Therefore,

if the left pan goes down, case (1) must be correct,

if a balance results, it must be case (2), and

if the right pan goes down, case (4) must be the one,

and all is revealed.

22. (From the University of Waterloo's 1988 "Euclid" Contest for grade 12 students in Canada)

If the lengths of the sides of a quadrilateral are 33, 47, 34, and 6, as in Figure 9, prove the diagonals are perpendicular.

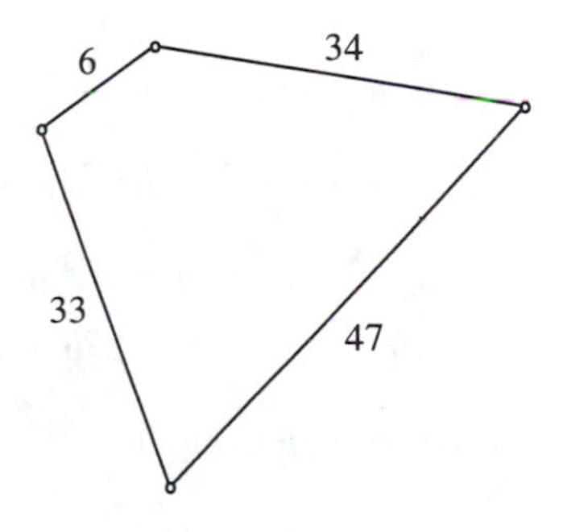

FIGURE 9

It would seem that in every approach to this problem the ensuing calculations soon uncover the secret property of the side-lengths 33, 47, 34, 6, namely that $33^2 + 34^2 = 47^2 + 6^2$ (= 2245). Thus it is a small step to the general conjecture that the diagonals of $ABCD$ are perpendicular if and only if

$$AB^2 + CD^2 = BC^2 + AD^2.$$

This is so easy to prove analytically that it is probably the simplest way to solve the problem.

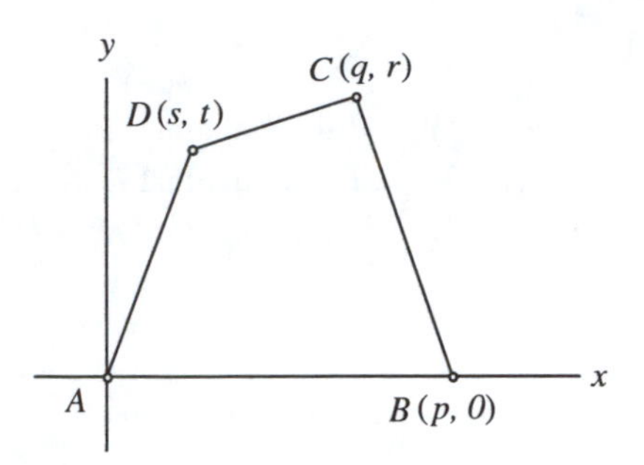

FIGURE 10

Let the coordinates of the vertices be $A(0,0)$, $B(p,0)$, $C(q,r)$, and $D(s,t)$ (Figure 10). Then AC and BD are perpendicular if and only if

$$\text{(the slope of AC)} \cdot \text{(the slope of BD)} = -1,$$

that is,

$$\frac{r}{q} \cdot \left(\frac{t}{s-p}\right) = -1,$$

or

$$q(p-s) = rt.$$

But this is precisely what $AB^2 + CD^2 = BC^2 + AD^2$ reduces to:

$$p^2 + [(s-q)^2 + (t-r)^2] = [(q-p)^2 + r^2] + (s^2 + t^2),$$
$$p^2 + s^2 - 2sq + q^2 + t^2 - 2tr + r^2 = q^2 - 2qp + p^2 + r^2 + s^2 + t^2,$$

in which all the perfect squares cancel, leaving just

$$2qp - 2sq = 2tr,$$

and the desired

$$q(p-s) = rt.$$

We observe that since the coordinates of C and D are unrestricted, the theorem applies not only to convex quadrilaterals but also to nonconvex and even self-intersecting quadrilaterals.

23. Without actually evaluating the integer $n = 5 \cdot 7^{34}$, prove that some digit occurs at least four times in its decimal representation.

The pigeonhole principle implies that every integer having more than 30 digits must have at least one of the ten decimal digits occur at least four times. Unfortunately this is no help here for the number n does not have more than 30 digits:

$$\begin{aligned}\log n &= \log 5 + 34 \cdot \log 7\\ &= .6990 + 34(.8451) \quad \text{to 4 places of decimals,}\\ &= .6990 + 28.7334\\ &= 29.4324\\ &< 30.\end{aligned}$$

However, at least this shows that n does have 30 digits. Consequently, if none of the ten digits were to occur more than three times, then, in order to make up the 30 digits, each of them would have to occur exactly three times. In this case, the sum of the digits of would be divisible by three, from which it follows that n itself would have to be divisible by 3. But this is not so, for we have modulo 3 that

$$n = 5 \cdot 7^{34} \equiv 2(1)^{34} = 2.$$

Thus some digit must occur at least four times in n.

(This problem was adapted from a problem for advanced 10–11 year old pupils in Czechoslovakia; in the original version, it was given that n has 30 digits and in a first part of the question it was asked whether n is divisible by three.)

24. (A Leningrad problem for pupils of age 14–16)

If positive integers A, B, C are such that

$$B \text{ divides } A^3, \quad C \text{ divides } B^3, \quad \text{and} \quad A \text{ divides } C^3,$$

prove that ABC divides $(A + B + C)^{13}$.

Since $C \mid B^3$, all the prime divisors of C are among the prime divisors of B; similarly, $B \mid A^3$ implies the prime divisors of B are also prime divisors of A. Hence any prime divisor of C is a prime divisor of A.

Conversely, $A \mid C^3$ implies that every prime divisor of A is a prime divisor of C. Thus A and C have the same set of prime divisors. Similarly, A and B have the same set of prime divisors, and it follows that A, B, C all have the same set of prime divisors.

Suppose the prime p occurs to the powers p^a, p^b, p^c, respectively, in the prime decompositions of A, B, C. Then

$$B \mid A^3 \text{ implies } b \leq 3a, \quad C \mid B^3 \text{ implies } c \leq 3b, \quad A \mid C^3 \text{ implies } a \leq 3c.$$

Thus we have

$$b \leq 3a,$$
$$c \leq 3b \leq 9a,$$

giving

$$b + c \leq 12a$$

and therefore

$$a + b + c \leq 13a.$$

Similarly,

$$a + b + c \leq 13b \quad \text{and} \quad a + b + c \leq 13c.$$

Thus, letting d denote the minimum value of (a, b, c), we have

$$a + b + c \leq 13d.$$

Now, the prime decomposition of ABC contains p to the power p^{a+b+c}, that is to say, to a power not exceeding p^{13d}. But clearly, p^d divides each of A, B, and C, and is therefore a divisor of $A + B + C$. Hence

$$(A + B + C)^{13} \text{ is divisible by } p^{13d},$$

and thus also by p^{a+b+c} (which is $\leq p^{13d}$). Since this holds for each prime divisor p of ABC, it follows that

$$ABC \text{ divides } (A + B + C)^{13}.$$

25. (From the Autumn 1991 Tournament of the Towns Competition)

The sequence $\{a_1, a_2, a_3, \ldots\}$ is defined by

$$a_0 = 9 \quad \text{and} \quad a_{k+1} = 3a_k^4 + 4a_k^3, \quad \text{for } k = 0, 1, 2, \ldots.$$

Prove that the decimal representation of a_{10} contains more than 1000 9's.

First, let us take a look at a little more of this sequence. To begin, we have

$$a_1 = 3 \cdot 9^4 + 4 \cdot 9^3 = 3 \cdot 6561 + 4 \cdot 729 = 19683 + 2916 = 22599.$$

Since a_2 promises to have 17 or 18 digits, the direct knowledge of a_0 and a_1 will have to suffice. Despite the meagerness of this evidence, it is enough to suggest the possibility that every a_k ends in a string of 9's. If this is true, then $a_k + 1$ would end in a string of 0's.

Following this line, suppose

$$a_k + 1 \text{ is divisible by } 10^n,$$

i.e.,

$$a_k + 1 = t \cdot 10^n \text{ for some positive integer } t,$$

and

$$a_k = t \cdot 10^n - 1.$$

The problem is to find out how this effects $a_{k+1} + 1$. We have

$$\begin{aligned} a_{k+1} + 1 &= 3a_k^4 + 4a_k^3 + 1 \\ &= 3(t \cdot 10^n - 1)^4 + 4(t \cdot 10^n - 1)^3 + 1 \\ &= 3(t^4 \cdot 10^{4n} - 4t^3 \cdot 10^{3n} + 6t^2 \cdot 10^{2n} - 4t \cdot 10^n + 1) \\ &\quad + 4(t^3 \cdot 10^{3n} - 3t^2 \cdot 10^{2n} + 3t \cdot 10^n - 1) + 1 \\ &= 3t^4 \cdot 10^{4n} - 8t^3 \cdot 10^{3n} + 6t^2 \cdot 10^{2n} \\ &= 10^{2n}(3t^4 \cdot 10^{2n} - 8t^3 \cdot 10^n + 6t^2). \end{aligned}$$

Thus $a^{k+1} + 1$ is divisible by 10^{2n}, and ends in (at least) twice as many 0's as $a_k + 1$; that is to say, a_{k+1} ends in (at least) twice as many 9's as a_k.

Since a_0 ends in one 9, that is, 2^0 of them, then a_1 ends in 2 of them and, in general, a_k ends in 2^k 9's. In particular, a_{10} ends in $2^{10} = 1024$ of them, and the conclusion follows.

26. (This is a slightly altered version of a problem from the Autumn 1991 Tournament of the Towns Competition. It is included here as an application of a result, relating to Napoleon's Theorem, that is established in the essay "The Nine-Point Circle and Coolidge's Theorem, The De Longchamps Point of a Triangle, Cantor's Theorem, and Napoleon's Theorem," which appears later in this volume.)

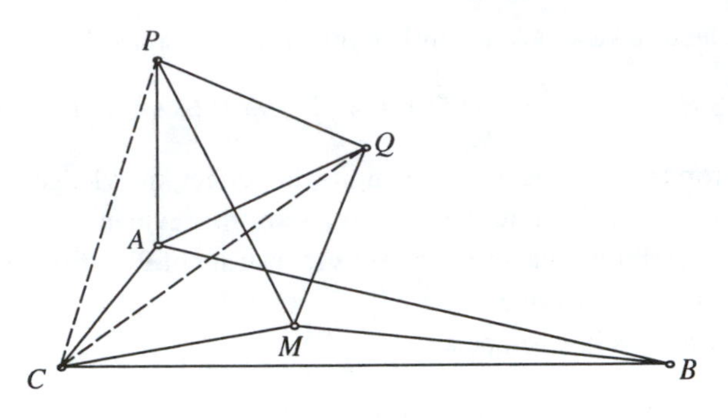

FIGURE 11

Let M be the centroid of $\triangle ABC$. Suppose MB is rotated about M through an angle of $120°$ to take B to the point P and that MC is rotated about M through $240°$ to take C to Q (Figure 11). Prove $\triangle APQ$ is equilateral.

As usual, let 1, ω, ω^2 denote the cube roots of unity. Recall that multiplication of a vector by ω rotates the vector through an angle of $120°$ without altering its magnitude. Hence the vector

$$\mathbf{MP} = \omega \cdot \mathbf{MB}, \quad \text{and similarly} \quad \mathbf{MQ} = \omega^2 \cdot \mathbf{MC}.$$

Thus the vector

$$\mathbf{CP} = \mathbf{CM} + \mathbf{MP} = \mathbf{CM} + \omega \cdot \mathbf{MB},$$

and

$$\mathbf{CQ} = \mathbf{CM} + \mathbf{MQ} = \mathbf{CM} + \omega^2 \cdot \mathbf{MC}.$$

Let C be taken as the origin of vectors. Now, recall that $\triangle XYZ$, determined by the vectors X, Y, Z from the origin, is equilateral if and only if

$$1 \cdot \mathbf{X} + \omega \cdot \mathbf{Y} + \omega^2 \cdot \mathbf{Z} = \mathbf{0},$$

where $X \to Y \to Z$ traces the triangle in the counterclockwise sense. For $\triangle APQ$, then, we need to show that

$$1 \cdot \mathbf{CA} + \omega \cdot \mathbf{CQ} + \omega^2 \cdot \mathbf{CP} = \mathbf{0}.$$

We have

$$\begin{aligned} &1 \cdot \mathbf{CA} + \omega \cdot \mathbf{CQ} + \omega^2 \cdot \mathbf{CP} \\ &\quad = \mathbf{CA} + \omega \cdot (\mathbf{CM} + \omega^2 \cdot \mathbf{MC}) + \omega^2 \cdot (\mathbf{CM} + \omega \cdot \mathbf{MB}) \end{aligned}$$

$$= \mathbf{CA} + \omega \cdot \mathbf{CM} + \mathbf{MC} + \omega^2 \cdot \mathbf{CM} + \mathbf{MB} \quad (\text{since } \omega^3 = 1)$$
$$= \mathbf{CA} + \mathbf{CM} \cdot (\omega - 1 + \omega^2) + (\mathbf{CB} - \mathbf{CM})$$
$$(\text{since } \mathbf{MC} = -\mathbf{CM} \text{ and } \mathbf{CM} + \mathbf{MB} = \mathbf{CB})$$
$$= \mathbf{CA} - 2 \cdot \mathbf{CM} + (\mathbf{CB} - \mathbf{CM}) \quad (\text{since } 1 + \omega + \omega^2 = 0)$$
$$= \mathbf{CA} + \mathbf{CB} - 3 \cdot \mathbf{CM}.$$

But since M is the centroid, $\mathbf{CM} = \frac{1}{3}(\mathbf{CA} + \mathbf{CB})$, giving the desired

$$1 \cdot \mathbf{CA} + \omega \cdot \mathbf{CQ} + \omega^2 \cdot \mathbf{CP} = \mathbf{CA} + \mathbf{CB} - 3 \cdot \mathbf{CM} = \mathbf{0}.$$

27. (From a 1987 Hungarian contest for 11-year-olds)

    How can a $3 \times 3 \times 3$ cube be cut up into 20 cubes (not necessarily the same size)?

I used to flatter myself that I would immediately be able to see through any problem that might be asked of an 11-year-old. I don't take anything for granted anymore!

Clearly a $1 \times 2 \times 2$ block can be quartered into unit cubes (Figure 12) and a $1 \times 3 \times 3$ block partitioned into nine unit cubes; also, a $2 \times 2 \times 2$ cube yields eight unit cubes. Thus it is easy to fritter away one's time trying out various schemes of cutting off a slice from the cube and further cutting the resulting pieces into a total of 20 cubes.

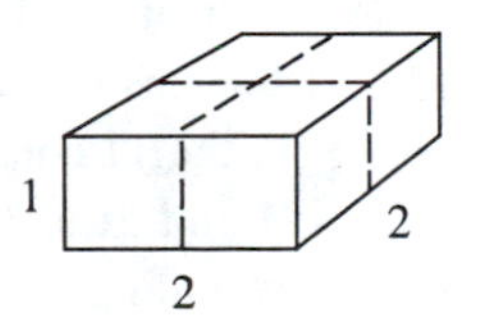

FIGURE 12

Surely one can't consider this problem without sometime having the thought that a $3 \times 3 \times 3$ cube can be divided into 27 unit cubes by cuts that trisect its edges. Since only 20 cubes are wanted, this doesn't seem to be much help and one might dismiss the idea out of hand. However, when you recall that eight unit cubes can be assembled into a single $2 \times 2 \times 2$ cube, the number of cubes drops from 27 to the desired 20. Thus it is clear that one should cut a solid $2 \times 2 \times 2$ cube out of one corner and then cut the remainder into 19 unit cubes with cuts that trisect the edges of the original cube.

28. (From the 1987 National Mathematics Competition in Yugoslavia)

The function f is defined on the positive integers as follows:

$$f(x) = \begin{cases} x - 10 & \text{for } x > 100 \\ f(f(x+11)) & \text{for } x \le 100 \end{cases}.$$

Prove that $f(x) = 91$ for all $x \le 100$.

This is an opportunity to use a "descending" induction.

(i) First observe that, for $n = 90, 91, \ldots, 100$, we have $n + 11 > 100$, in which case $f(n+11) = (n+11) - 10 = n + 1$ and therefore, since n itself is ≤ 100, that

$$f(n) = f(f(n+11)) = f(n+1).$$

It follows, then, that

$$f(90) = f(91) = \cdots = f(100) = f(101) = 91 \quad (\text{since } 101 > 100).$$

(ii) Now consider the values of n in decreasing order $100, 99, 98, \ldots, 2, 1$. If it happens, for some value of $n \le 89$, that $f(n+11) = 91$, then $f(n) = f(f(n+11))$ would equal $f(91)$, which we have seen above is also 91. Therefore, in determining the value of $f(n)$, for $n \le 89$, look back at the value of $f(n+11)$; if it is 91, then so is $f(n) = 91$. But more importantly, if the value of the function is 91 **for each of the 11 consecutive larger values of n**, then not only will $f(n)$ be 91 but this will extend to $n - 1$ the property that the value of the function is 91 for the 11 consecutive larger values, which similarly implies that $f(n-1) = 91$ and extends the "consecutive" property to $n - 2$, giving $f(n-2) = 91$ and extending the property.... Since this is the case for $n = 89$, it follows by a descending induction that $f(n) = 91$ for all $n \le 89$, completing the proof.

29. Now for a beautiful problem from the Senior Division of the 1989 Tournament of the Towns.

On an $s \times s$ square piece of paper, whose edges run horizontally and vertically, there are a number of inkblots, each of area ≤ 1 (Figure 13). If each horizontal line and each vertical line across the paper makes contact with at most one inkblot, prove that the total area of the inkblots, however many of them there might be, does not exceed s.

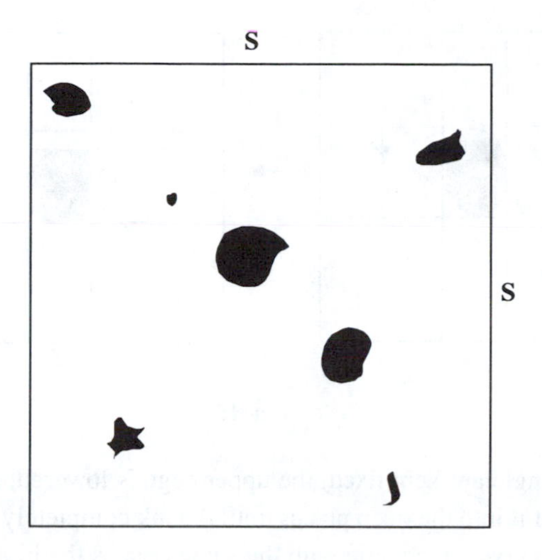

FIGURE 13

In view of the property concerning horizontal and vertical lines, its hard to resist enclosing each blot in a rectangle of minimum size with tangent lines that are horizontal and vertical (Figure 14). Since inkblots are generally of irregular shape, there is probably a lot of space in these rectangles around the outside of the blot.

Suppose there are n blots. Also suppose that the ink has neither dried nor soaked into the paper and that, while the bottom edge and the sides of each

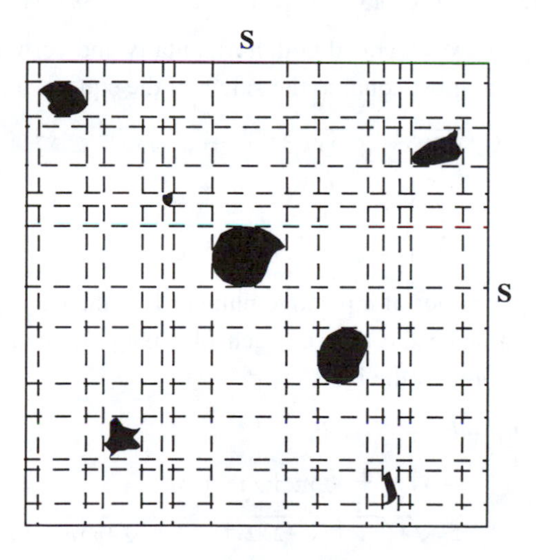

FIGURE 14

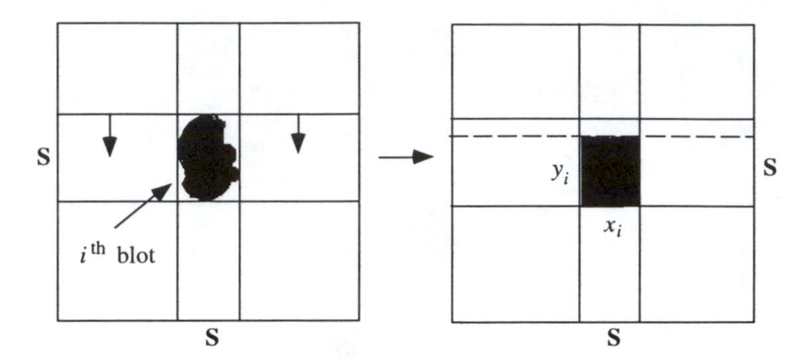

FIGURE 15

enclosing rectangle are kept fixed, the upper edge is lowered, squeegeeing all the ink ahead of it into the open places until the ink completely fills the shrinking rectangle to give a rectangle with the same area as the blot (Figure 15). If the horizontal and vertical dimensions of the final rectangle obtained from the ith blot are x_i and y_i, respectively, then the area b_i of the blot is simply $x_i y_i$ (Figure 15).

Now, the Cauchy inequality yields

$$(u_1v_1 + u_2v_2 + \cdots + u_nv_n)^2 \le (u_1^2 + u_2^2 + \cdots + u_n^2)(v_1^2 + v_2^2 + \cdots + v_n^2)$$

for positive numbers $u_1, u_2, \ldots, u_n$ and $v_1, v_2, \ldots, v_n$.

Hence for $u_i = \sqrt{x_i}$ and $v_i = \sqrt{y_i}$, we obtain

$$(\sqrt{x_1y_1} + \sqrt{x_2y_2} + \cdots + \sqrt{x_ny_n})^2 \le (x_1 + x_2 + \cdots + x_n)(y_1 + y_2 + \cdots + y_n).$$

Since the rectangles are separated both horizontally and vertically, each factor on the right side is at most equal to the side s of the square, and we have

$$(\sqrt{x_1y_1} + \sqrt{x_2y_2} + \cdots + \sqrt{x_ny_n})^2 \le s^2$$

and

$$\sqrt{x_1y_1} + \sqrt{x_2y_2} + \cdots + \sqrt{x_ny_n} \le s.$$

Now, the square root of a positive number less than 1 is greater than the number itself, and since the area of each blot is ≤ 1, then $x_iy_i \le \sqrt{x_iy_i}$. Therefore the total area of the blots is

$$\begin{aligned} b_1 + b_2 + \cdots + b_n & \\ &= x_1y_1 + x_2y_2 + \cdots + x_ny_n \\ &\le \sqrt{x_1y_1} + \sqrt{x_2y_2} + \cdots + \sqrt{x_ny_n} \\ &\le s. \end{aligned}$$

30. (From the 1990 Tournament of the Towns—Years 8, 9, 10)

Let

$$a = \cfrac{1}{2 + \cfrac{1}{3 + \cfrac{1}{\begin{matrix} \vdots \\ \hline 97 + \cfrac{1}{98 + \cfrac{1}{99}} \end{matrix}}}}$$

and

$$b = \cfrac{1}{2 + \cfrac{1}{3 + \cfrac{1}{\begin{matrix} \vdots \\ \hline 97 + \cfrac{1}{98 + \cfrac{1}{99 + \cfrac{1}{100}}} \end{matrix}}}}$$

Prove that $|a - b| < \dfrac{1}{99!\,100!}$.

There doesn't seem to be much hope of doing anything with a and b "from the top," so let's see what can be done from the bottom. To this end, let

$$x_n = n + \cfrac{1}{n + 1 + \cfrac{1}{n + 2 + \cfrac{1}{\begin{matrix} \vdots \\ \hline \cfrac{1}{98 + \cfrac{1}{99}} \end{matrix}}}},$$

$$y_n = n + \cfrac{1}{n+1+\cfrac{1}{n+2+\cfrac{1}{\begin{array}{c}\vdots\\ \hline 1\\ \hline 98+\cfrac{1}{99+\cfrac{1}{100}}\end{array}}}}.$$

In this case, $a = x_1 - 1$ and $b = y_1 - 1$, making $a - b = x_1 - y_1$. Thus we want to show that $|x_1 - y_1| < \frac{1}{99!\,100!}$.

Clearly

$$x_n = n + \frac{1}{x_{n+1}} \quad \text{and} \quad y_n = n + \frac{1}{y_{n+1}}.$$

Hence

$$x_n - y_n = \frac{1}{x_{n+1}} - \frac{1}{y_{n+1}} = \frac{y_{n+1} - x_{n+1}}{x_{n+1}y_{n+1}},$$

and so

$$|x_n - y_n| = \frac{1}{x_{n+1}y_{n+1}} \cdot |x_{n+1} - y_{n+1}|.$$

Now, for $n \le 97$, each of x_{n+1} and y_{n+1} exceeds $n+1$, and since $x_{99} = 99$ and $y_{99} = 99 + \frac{1}{100} > 99$, then the product $x_{n+1}y_{n+1} > (n+1)^2$ for all $n = 1, 2, \ldots, 98$. Thus $\frac{1}{x_{n+1}y_{n+1}} < \frac{1}{(n+1)^2}$ for $n \le 98$ and

$$|x_n - y_n| < \frac{1}{(n+1)^2} \cdot |x_{n+1} - y_{n+1}|.$$

Finally, then,

$$\begin{aligned}
|x_1 - y_1| &< \frac{1}{2^2}|x_2 - y_2| \\
&< \frac{1}{2^2 \cdot 3^2}|x_3 - y_3| \quad \left(= \frac{1}{(3!)^2}|x_3 - y_3|\right) \\
&< \frac{1}{(4!)^2}|x_4 - y_4| \\
&\ \ \vdots \\
&< \frac{1}{(99!)^2}|x_{99} - y_{99}|,
\end{aligned}$$

$$= \frac{1}{(99!)^2}\left|99 - \left(99 + \frac{1}{100}\right)\right|$$
$$= \frac{1}{99!\,100!}.$$

31. (From the 1990 Tournament of the Towns—Years 8, 9, 10)

$X = \{x_1, x_2, x_3, \ldots\}$ is a sequence of real numbers such that

$$x_{n+1} = |x_n| - x_{n-1} \quad \text{for } n = 2, 3, \ldots.$$

Prove that X has period 9.

First, observe that X cannot consist only of positive terms:

if a and b are consecutive positive terms, then X proceeds

$$a, b, b - a, \ldots,$$

at which point either $b - a$ itself fails to be positive, or the term following it is $|b - a| - b = (b - a) - b = -a$, a negative number.

Thus X must contain a nonpositive term no later than its fourth term.

From a nonpositive term $-a$, $a \geq 0$, X must proceed either with another nonpositive term $-b$ ($b \geq 0$) or with a positive term c.

(i) Suppose X proceeds $-a, -b, \ldots$: in this case, succeeding terms are

$$-a, -b, b + a, 2b + a, b, -b - a, a, 2a + b, a + b, -a, -b, \ldots,$$

which repeats with period 9.

(ii) Suppose X proceeds $-a, c, \ldots$: then X continues

$$-a, c, c + a, a, -c, c - a,$$

at which point succeeding terms depend on the value of $c - a$:

if $c - a \geq 0$, X continues $\quad 2c - a, c, -c + a, -a, c, \ldots$
if $c - a < 0$, X continues $\quad a, 2a - c, a - c, -a, c, \ldots$
(since $c - a < 0$, then $a - c > 0$, and so $2a - c > 0$),

each of which shows that the sequence repeats with period 9. It follows, then, that X repeats with period 9 from the first nonpositive term, which must occur somewhere among its first four terms.

Now, the recursion also allows us to calculate the terms preceding a nonpositive term $-a$:

$$x_{n-1} = |x_n| - x_{n+1}.$$

Therefore any terms that precede $-a$ occur again just ahead of the second appearance of $-a$, implying X repeats with period 9 from its very first term.

$$x_1, \ldots, -a, k, \ldots, x_1, \ldots, -a, k, \ldots$$

32. It is a pleasure to thank my colleagues Frank Zorzitto and Lee Dickey for telling me about the following gem.

Prove that $|\sin\alpha - \sin\beta| < |\alpha - \beta|$.

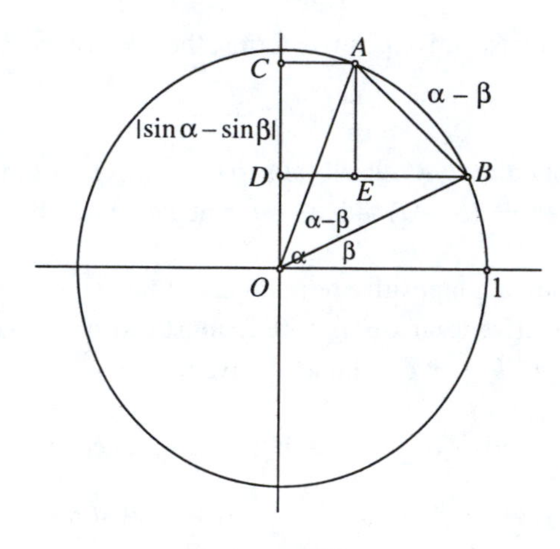

FIGURE 16

In Figure 16, $\sin\alpha$ and $\sin\beta$ are given by the ordinates of A and B in the unit circle, that is, by OC and OD, and so

$$|\sin\alpha - \sin\beta| = CD.$$

Also, $|\alpha - \beta|$ is given by the arc AB subtended at the circumference of the unit circle by the angle $\alpha - \beta$.

Now, if AE is perpendicular to BD, then $CD = AE$, and we have

$$\begin{aligned} |\alpha - \beta| &= \text{arc } AB \\ &> \text{hypotenuse } AB \text{ in right triangle } ABE \\ &> \text{leg } AE \\ &= |\sin\alpha - \sin\beta|. \end{aligned}$$

SECTION 17

Two Challenging Problems in Combinatorics

1. Problem 1

   How many sequences of 0's, 1's, 2's, and 3's are there, each with a total of n symbols, consisting of an even number of 0's, an odd number of 1's, and any number of 2's and 3's?

This is a great opportunity to show off a brilliant twist on ordinary generating functions called the *exponential generating function*. In case you are not familiar with this topic, let us begin with a brief introduction to it.

(a) An Introduction to Exponential Generating Functions Consider the number of ways a total of 12 0's and 1's can be arranged in a row. In the case when 5 0's and 7 1's are used, it is easy to see that the number of arrangements is $\binom{12}{5}$: simply pick 5 of the 12 places in the row for the 0's, and put the 1's in the other places. Now, the following observation establishes a crucial relation between this number and the product $\frac{x^5}{5!} \cdot \frac{x^{12}}{7!}$:

$$\frac{x^5}{5!} \cdot \frac{x^7}{7!} = \frac{x^{12}}{5!7!} = \frac{12!}{5!7!} \cdot \frac{x^{12}}{12!} = \binom{12}{5} \frac{x^{12}}{12!}.$$

That is to say, the number of ways of arranging 5 0's and 7 1's in a row is given by the resulting coefficient when the product $\frac{x^5}{5!} \cdot \frac{x^7}{7!}$ is expressed as a term in $\frac{x^{12}}{12!}$. The same is true in general:

$$\frac{x^r}{r!} \cdot \frac{x^s}{s!} = \frac{x^{r+s}}{r!s!} = \frac{(r+s)!}{r!s!} \cdot \frac{x^{r+s}}{(r+s)!} = \binom{r+s}{r} \frac{x^{r+s}}{(r+s)!},$$

in which the coefficient $\binom{r+s}{s}$ of the term $\frac{x^{r+s}}{(r+s)!}$ is the number of ways of arranging r 0's and s 1's in a row. Clearly the infinite sum

$$f(x) = \sum \frac{x^r}{r!} \cdot \frac{x^s}{s!}, \quad \text{taken over all } r, s \geq 0,$$

contains all possible products, in particular the products for sequences of 12 symbols,

$$\frac{x^0}{0!}\cdot\frac{x^{12}}{12!}, \frac{x^1}{1!}\cdot\frac{x^{11}}{11!}, \ldots, \frac{x^{12}}{12!}\cdot\frac{x^0}{0!}$$

Therefore the total number of ways of arranging a total of 12 0's and 1's in a row is given by the sum of the coefficients that are obtained by expressing these products as terms in $\frac{x^{12}}{12!}$:

$$\frac{x^0}{0!}\cdot\frac{x^{12}}{12!} \text{ gives } \binom{12}{0}, \quad \frac{x^1}{1!}\cdot\frac{x^{11}}{11!} \text{ gives } \binom{12}{1}, \quad \text{and so forth.}$$

If we had to deal with each term separately and add up the results, we wouldn't be any further ahead with this approach. But, as we shall see, all the cases can be handled in one painless operation.

Note that $f(x)$ is simply the square of e^x:

$$\begin{aligned} f(x) &= \sum \frac{x^r}{r!}\cdot\frac{x^s}{s!} \\ &= \left(1+\frac{x}{1!}+\frac{x^2}{2!}+\cdots\right)\left(1+\frac{x}{1!}+\frac{x^2}{2!}+\cdots\right) \\ &= e^{2x} \\ &= 1+\frac{(2x)}{1!}+\frac{(2x)^2}{2!}+\cdots+\frac{(2x)^{12}}{12!}+\cdots+\frac{(2x)^n}{n!}+\cdots. \end{aligned}$$

Thus, when all the products are expressed in terms of $\frac{x^{12}}{12!}$ and added together, the final result is $2^{12}\cdot\frac{x^{12}}{12!}$, implying there are 2^{12} ways of arranging a row of 12 0's and 1's. I expect you must be wondering whether this is really getting us anywhere, for we all knew this was the answer before we started (there are two choices, either a 0 or a 1, for each of the 12 places in the row). However, it is clear that exponential generating functions do have the worthwhile benefit of automatically taking care of the number of ways the various numbers of 0's and 1's can be mixed together in the row and, as we shall see, like all generating functions, they have marvellous flexibility.

Before considering another example, we note that the factor $\frac{x^r}{r!}$, corresponding to the use of r 0's in the row, was taken from the first bracket in $f(x)$, and that the factor $\frac{x^s}{s!}$, indicating the selection of s 1's for the row, came from the second bracket:

$$f(x) = \left(1+\frac{x}{1!}+\frac{x^2}{2!}+\cdots\right)\left(1+\frac{x}{1!}+\frac{x^2}{2!}+\cdots\right) = \cdots+\frac{x^r}{r!}\cdot\frac{x^s}{s!}+\cdots.$$

Thus this first bracket is considered to be the exponential generating function for the ways of picking the 0's and the second one the function for the ways of picking the 1's; the fact that these functions are the same is coincidental.

Suppose we wanted never to use exactly one 0 in the row. In order to eliminate this option, all we need to do is drop the term $\frac{x}{1!}$ from the generating function for the 0's. Thus the generating function for such constrained sequences is

$$\begin{aligned} f(x) &= \left(1 + \frac{x^2}{2!} + \frac{x^3}{3!} + \cdots\right)\left(1 + \frac{x}{1!} + \frac{x^2}{2!} + \cdots\right) \\ &= (e^x - x)e^x \\ &= e^{2x} - xe^x. \end{aligned}$$

Denoting the coefficient of x^n in a function $g(x)$ with square brackets, $[x^n]g(x)$, the number of such sequences of 12 symbols, then, is

$$\begin{aligned} \left[\frac{x^{12}}{12!}\right] f(x) &= \left[\frac{x^{12}}{12!}\right](e^{2x} - xe^x) \\ &= \left[\frac{x^{12}}{12!}\right] e^{2x} - \left[\frac{x^{12}}{12!}\right] xe^x \\ &= 2^{12} - \left[\frac{x^{12}}{12!}\right] x\left(1 + \frac{x}{1!} + \cdots + \frac{x^{11}}{11!} + \cdots\right) \\ &= 2^{12} - \left[\frac{x^{12}}{12!}\right]\left(x + \frac{x^2}{1!} + \cdots + \frac{x^{12}}{11!} + \cdots\right) \\ &= 2^{12} - \left[\frac{x^{12}}{12!}\right]\left(x + \frac{x^2}{1!} + \cdots + 12 \cdot \frac{x^{12}}{12!} + \cdots\right) \\ &= 2^{12} - 12. \end{aligned}$$

Again, this might make you wonder about the whole business, since the -12 merely deducts the $\binom{12}{1}$ sequences in which exactly one 0 occurs. Admittedly our examples have concerned inconsequential benefits, but they have served to illustrate some basic ideas about exponential generating functions and we are now in a position to proceed with a beautiful solution to the problem at hand.

(b) The Solution The exponential generating function for the ways of selecting an even number of 0's is

$$1 + \frac{x^2}{2!} + \frac{x^4}{4!} + \cdots,$$

and the function for an odd number of 1's is

$$x+\frac{x^3}{3!}+\frac{x^5}{5!}+\cdots,$$

while the generating function for the unconstrained choices of 2's and 3's is, in each case, the unmodified series for e^x. Hence the exponential generating function for the sequences in question is

$$f(x)=\left(1+\frac{x^2}{2!}+\frac{x^4}{4!}\right)\cdot\left(x+\frac{x^3}{3!}+\frac{x^5}{5!}+\cdots\right)\cdot e^x\cdot e^x.$$

Now, the first two factors in this function are the results of "bisecting" the exponential series:

$$e^x=1+\frac{x}{1!}+\frac{x^2}{2!}+\frac{x^3}{3!}+\cdots,$$

and

$$e^{-x}=1-\frac{x}{1!}+\frac{x^2}{2!}-\frac{x^3}{3!}+-\cdots;$$

hence

$$\frac{1}{2}(e^x+e^{-x})=1+\frac{x^2}{2!}+\frac{x^4}{4!}+\cdots$$

and

$$\frac{1}{2}(e^x-e^{-x})=x+\frac{x^3}{3!}+\frac{x^5}{5!}+\cdots.$$

Thus

$$\begin{aligned}f(x)&=\frac{1}{2}(e^x+e^{-x})\cdot\frac{1}{2}(e^x-e^{-x})\cdot e^x\cdot e^x\\&=\frac{1}{4}(e^{2x}-e^{-2x})\cdot e^{2x}\\&=\frac{1}{4}(e^{4x}-1),\end{aligned}$$

and the desired number of sequences is simply the coefficient of $\frac{x^n}{n!}$ in this function, that is,

$$\left[\frac{x^n}{n!}\right]\frac{1}{4}(e^{4x}-1)=\frac{1}{4}\cdot 4^n=4^{n-1}.$$

Exercise: Prove that the number of ways of assigning 26 people to three rooms, 1, 2, and 3, so that no room remains empty is more than $2\frac{1}{2}$ *trillion.*

2. Terquem's Problem

This problem was considered in 1839 by Olry Terquem (1782–1862).

Let $\alpha = \{a_1, a_2, \ldots, a_k\}$ be a subset of the first n positive integers which has been arranged in increasing order:

$$1 \le a_1 < a_2 < \cdots < a_k \le n.$$

(i) How many such α are there that begin with an odd number and thereafter alternate in parity:

$$\alpha = \{\text{odd, even, odd, even, } \ldots\}?$$

For convenience, include the empty set $\emptyset$ in the count.

(ii) How many α are there of size k?

(i) Let $t(n)$ denote the total number of such α from $\{1, 2, \ldots, n\}$. Then

$$\begin{aligned}
t(1) &= 2 \quad \text{for} \quad \alpha = \emptyset, \{1\};\\
t(2) &= 3 \quad \text{for} \quad \alpha = \emptyset, \{1\}, \{1,2\};\\
t(3) &= 5 \quad \text{for} \quad \alpha = \emptyset, \{1\}, \{1,2\}, \{3\}, \{1,2,3\};\\
t(4) &= 8 \quad \text{for} \quad \alpha = \emptyset, \{1\}, \{1,2\}, \{3\}, \{1,2,3\}, \{1,4\},\\
& \quad \text{for} \quad \{3,4\}, \{1,2,3,4\},
\end{aligned}$$

strongly suggesting that $t(n) = f_{n+2}$, the $(n+2)$th term of the Fibonacci sequence

$$\{f_n\} = \{1, 1, 2, 3, 5, 8, 13, \ldots\}.$$

As we shall see, this is confirmed by observing that $t(n)$ satisfies the Fibonacci recurrence relation

$$t(n) = t(n-1) + t(n-2).$$

Clearly

$$\begin{aligned}
t(n) &= (\text{the number of } \alpha \text{ which do } \textbf{not} \text{ contain } n)\\
&\quad + (\text{the number of } \alpha \text{ which } \textbf{do} \text{ contain } n)\\
&= t(n-1) + (\text{the number of } \alpha \text{ which } \textbf{do} \text{ contain } n).
\end{aligned}$$

It remains to show that

$$(\text{the number of } \alpha \text{ which } \textbf{do} \text{ contain } n) = t(n-2).$$

For definiteness, suppose n is even. Now, there are two kinds of α that contain n (and therefore end in n), and two kinds of ordered alternating subsets β that are counted by $t(n-2)$:

(i) $\alpha = \{a_1, a_2, \ldots, \text{even}, n-1, n\}$, ending in $n-1, n$ (an odd-even pair),
(ii) $\alpha = \{a_1, a_2, \ldots, \text{odd}, n\}$, where the final odd number is $\leq n-3$,
(iii) $\beta = \{a_1, a_2, \ldots, \text{even}\}$, where the final even number is $\leq n-2$,
(iv) $\beta = \{a_1, a_2, \ldots, \text{odd}\}$, where the final odd number is $\leq n-3$.

Clearly, deleting the $n-1$ and n from an α of type (i) gives a β of type (iii), and conversely; and, deleting the n from an α of type (ii) gives a β of type (iv), and conversely. The conclusion follows.

Now for the more challenging and interesting part of the problem.

(ii) We want to count the number of $\alpha = \{a_1, a_2, \ldots, a_k\}$ for a given value of k. Because the integers in α alternate in parity, the $k-1$ differences

$$a_2 - a_1, a_3 - a_2, a_4 - a_3, \ldots, a_k - a_{k-1}$$

are all odd. Since a_1 is also odd, any k-tuple of positive odd integers

$$d = \{d_1, d_2, \ldots, d_k\}$$

gives rise to an increasing, alternating subset

$$\gamma = \{d_1, d_1 + d_2, d_1 + d_2 + d_3, \ldots, d_1 + d_2 + \cdots + d_k\},$$

which would qualify as an α if and only if its final integer

$$d_1 + d_2 + \cdots + d_k \leq n.$$

That is to say, there is a 1-1 correspondence between our subsets α and the k-tuples d of positive odd integers whose components do not add up to more than n.

Now, these d are easily counted. Clearly, they are generated by the function

$$\begin{aligned} f(x) &= (x + x^3 + x^5 + \cdots)^k \\ &= \cdots + x^{d_1 + d_2 + \cdots + d_k} + \cdots, \\ &= \cdots + c_r x^r + \cdots, \end{aligned}$$

where c_r is the number of k-tuples d whose components add up to r. The total number N of k-tuples d, then, is the **sum** of the coefficients of all the terms up to x^n:

$$N = c_0 + c_1 + c_2 + \cdots + c_n.$$

How to add up these coefficients is an interesting challenge; can you see a neat way of doing it? The idea is to incorporate them all into the coefficient of x^n by multiplying $f(x)$ by $(1 + x + x^2 + x^3 + \cdots)$:

$$\begin{aligned} &f(x) \cdot (1 + x + x^2 + x^3 + \cdots) \\ &\quad = (x + x^3 + x^5 + \cdots)^k (1 + x + x^2 + x^3 + \cdots) \\ &\quad = (c_0 x^0 + c_1 x + c_2 x^2 + \cdots)(1 + x + x^2 + x^3 + \cdots) \\ &\quad = \cdots + (c_0 + c_1 + c_2 + \cdots + c_n) x^n + \cdots. \end{aligned}$$

Thus the number N of k-tuples d, which is also the desired number of subsets α of size k, is simply

$$N = [x^n] f(x) \cdot (1 + x + x^2 + x^3 + \cdots).$$

It remains only to carry out the straightforward calculations.

Recalling that $1 + x + x^2 + x^3 + \cdots = (1 - x)^{-1}$, we have

$$\begin{aligned} N &= [x^n](x + x^3 + x^5 + \cdots)^k (1 - x)^{-1} \\ &= [x^n] x^k (1 + x^2 + x^4 + \cdots)^k (1 - x)^{-1} \\ &= [x^{n-k}][(1 - x^2)^{-1}]^k (1 - x)^{-1} \\ &= [x^{n-k}](1 - x^2)^{-k} (1 - x)^{-1} [(1 + x)^{-1}(1 + x)] \\ &\qquad\qquad\qquad\qquad (\text{since } [(1 + x)^{-1}(1 + x)] = 1) \\ &= [x^{n-k}](1 - x^2)^{-(k+1)}(1 + x). \end{aligned}$$

Recalling that

$$(1 - y)^{-m} = \sum_{i \geq 0} \binom{m + i - 1}{i} y^i,$$

we have

$$N = [x^{n-k}](1 + x) \sum_{i \geq 0} \binom{k + 1 + i - 1}{i} (x^2)^i$$

and since $\binom{n}{r} = \binom{n}{n-r}$,

$$N = [x^{n-k}] \sum_{i \geq 0} \binom{k+i}{k} x^{2i} + [x^{n-k}] \sum_{i \geq 0} \binom{k+i}{k} x^{2i+1}.$$

For $n-k = 2i$, we have $i = \frac{n-k}{2}$, and for $n-k = 2i+1$, we have $i = \frac{n-k-1}{2}$. Thus in all cases $i = \lfloor \frac{n-k}{2} \rfloor$, the integer part of $\frac{n-k}{2}$, and we have finally that

$$N = \binom{k + \lfloor \frac{n-k}{2} \rfloor}{k}.$$

SECTION 18

An Unused Problem from the 1988 International Olympiad

Proposed by The Federal Republic of Germany

Observe that the 20 ordered partitions of 6 which contain one or more parts equal to 2 are

$$\begin{array}{rrrr} 2+2+2, & 2+2+1+1, & 2+1+2+1, & 2+1+1+2, \\ 1+2+2+1, & 1+2+1+2, & 1+1+2+2, & 2+4, \\ 4+2, & 2+1+3, & 2+3+1, & 1+2+3, \\ 3+2+1, & 1+3+2, & 3+1+2, & 2+1+1+1+1, \\ 1+2+1+1+1, & 1+1+2+1+1, & 1+1+1+2+1, & 1+1+1+1+2, \end{array}$$

containing a total of 28 parts equal to 2. How many parts equal to k are there in all the ordered partitions of n?

Our first solution is based on the brilliant approach given in the 250-page report on this olympiad—*An Olympiad Down Under*, published by the Australian Mathematics Foundation. The second solution is a beautiful application of generating functions that I hope you will also find exciting.

Solution 1 It is not unusual to represent an ordered partition of n as a row of n dots suitably sectioned by vertical dividers; for example, in the case of $n = 6$, the ordered partitions

$$3+1+2 \quad \text{and} \quad 1+2+3$$

are represented respectively by

$$\bullet\ \bullet\ \bullet \,|\, \bullet \,|\, \bullet\ \bullet \quad \text{and} \quad \bullet \,|\, \bullet\ \bullet \,|\, \bullet\ \bullet\ \bullet\,.$$

Now, in a row of dots which represents a partition of 6, a part equal to 2 might occur as dots 1 and 2, or dots 2 and 3, ..., or dots 5 and 6:

$$(1,2), (2,3), (3,4), (4,5), (5,6).$$

Observe that the number of partitions in which a part equal to 2 occupies positions (2, 3) is four:

$$\begin{array}{ll}
\bullet \mid \bullet \ \bullet \mid \bullet \ \bullet \ \bullet & (1+2+3) \\
\bullet \mid \bullet \ \bullet \mid \bullet \mid \bullet \ \bullet & (1+2+1+2) \\
\bullet \mid \bullet \ \bullet \mid \bullet \ \bullet \mid \bullet & (1+2+2+1) \\
\bullet \mid \bullet \ \bullet \mid \bullet \mid \bullet \mid \bullet & (1+2+1+1+1).
\end{array}$$

There are also four partitions in which a part equal to 2 occupies the positions (3, 4):

$$\begin{array}{ll}
\bullet \ \bullet \mid \bullet \ \bullet \mid \bullet \ \bullet & (2+2+2) \\
\bullet \ \bullet \mid \bullet \ \bullet \mid \bullet \mid \bullet & (2+2+1+1) \\
\bullet \mid \bullet \mid \bullet \ \bullet \mid \bullet \ \bullet & (1+1+2+2) \\
\bullet \mid \bullet \mid \bullet \ \bullet \mid \bullet \mid \bullet & (1+1+2+1+1),
\end{array}$$

and it is easy to check that there are also four partitions in which a part equal to 2 occurs in positions (4, 5). The reason is that, besides the two dividers which define the specified part equal to 2, in each case there are two remaining places where a divider may be inserted or which may be left blank, providing $2^2 = 4$ ways of completing a partition having the specified 2 in the prescribed position.

Now, there is no need to enter a divider at either end of the row; dividers only separate sections when they are entered between two dots. Thus two dividers are necessary to mark off a specified part in the interior of the row but only one divider is required for a part which occurs at the beginning or the end of the row. In the latter case, then, there is an extra space into which a divider may or may not be placed in completing the partition, thus doubling the number of options. Hence, for example, there are $2 \cdot 4 = 8$ partitions of 6 which begin with 2:

$$\begin{array}{ll}
\bullet \ \bullet \mid \bullet \ \bullet \ \bullet \ \bullet & (2+4) \\
\bullet \ \bullet \mid \bullet \mid \bullet \ \bullet \ \bullet & (2+1+3) \\
\bullet \ \bullet \mid \bullet \ \bullet \mid \bullet \ \bullet & (2+2+2) \\
\bullet \ \bullet \mid \bullet \ \bullet \ \bullet \mid \bullet & (2+3+1) \\
\bullet \ \bullet \mid \bullet \mid \bullet \mid \bullet \ \bullet & (2+1+1+2) \\
\bullet \ \bullet \mid \bullet \mid \bullet \ \bullet \mid \bullet & (2+1+2+1) \\
\bullet \ \bullet \mid \bullet \ \bullet \mid \bullet \mid \bullet & (2+2+1+1) \\
\bullet \ \bullet \mid \bullet \mid \bullet \mid \bullet \mid \bullet & (2+1+1+1+1).
\end{array}$$

Hence, with the eight partitions that end with 2, the total number of parts equal to 2 in the ordered partitions of 6 is

$$T(2,6) = 8 + 4 + 4 + 4 + 8 = 28.$$

The general case is exactly the same. In a row of n dots, an interior part equal to k can begin with the ith dot, where $i = 2, 3, \ldots, n-k$ (a part starting with the $(n-k)$th dot extends over the k dots numbered $n-k, n-k+1, \ldots,$ $n-k+(k-1) = n-1$, and thus lies completely in the interior of the row).

$$\underbrace{\bullet \; \bullet \; \cdots \; \bullet}_{i-1} \Big| \underbrace{\bullet \; \bullet \; \cdots \; \bullet}_{k} \Big| \underbrace{\bullet \; \bullet \; \cdots \; \bullet}_{n-k-i+1}$$

In each case, the number of interior places for additional dividers is

$$[(i-1)-1] + [(n-k-i+1)-1] = n-k-2,$$

giving 2^{n-k-2} ordered partitions of n which hold a part equal to k in the prescribed position. The $(n-k-1)$ possible interior positions thus account for a total of $(n-k-1) \cdot 2^{n-k-2}$ parts equal to k, and since there are twice as many partitions that begin or end with k as for an interior position, the total number of parts equal to k is

$$\begin{aligned} T(k,n) &= 2^{n-k-1} + (n-k-1) \cdot 2^{n-k-2} + 2^{n-k-1} \\ &= 2^{n-k} + (n-k-1) \cdot 2^{n-k-2}. \end{aligned}$$

For $n = 6, k = 2$, this gives $2^4 + 3 \cdot 2^2 = 16 + 12 = 28$.

Solution 2 Consider the series $f(u,x)$ in which the term x^k is distinguished with a coefficient u:

$$f(u,x) = x + x^2 + \cdots + x^{k-1} + ux^k + x^{k+1} + \cdots.$$

In the product

$$\begin{aligned} [f(u,x)]^m &= (x + x^2 + \cdots + ux^k + \cdots)^m \\ &= \cdots + \underbrace{x^i \cdot x^j \cdot ux^k \cdot x^s \cdot ux^k \cdot \cdots \cdot x^r}_{\ldots m \text{ factors} \ldots} + \cdots \\ &= \cdots + a_{t,n} u^t x^n + \cdots, \end{aligned}$$

the coefficient $a_{t,n}$ is the number of ways an ordered sum of n can be obtained by adding m positive integers, t of which are the integer k, that is, $a_{t,n}$ is the number of m-part ordered partitions of n which contain t k's. Now, we wish to consider all ordered partitions of n, however many parts they might have. Accordingly, in the generating function

$$\begin{aligned} F(u,x) &= [f(u,x)]^0 + [f(u,x)]^1 + [f(u,x)]^2 + [f(u,x)]^3 + \cdots \\ &= 1 + (x + x^2 + \cdots + ux^k + \cdots) + (x + x^2 + \cdots + ux^k + \cdots)^2 \\ &\quad + (x + x^2 + \cdots + ux^k + \cdots)^3 + \cdots \\ &= \cdots + c_{t,n}u^t x^n + \cdots, \end{aligned}$$

the simplified coefficient $c_{t,n}$ is the total number of ordered partitions of n which contain t parts equal to k, counting partitions of all lengths (it does no harm to include all powers of $f(u,x)$ in this generating function, for there is no term in x^n in any power of $f(u,x)$ whose exponent is greater than n). If $n = qk + r$, where $0 \le r < k$, no partition of n can contain more than q parts equal to k, and gathering all the terms in x^n, we obtain

$$F(u,x) = \cdots + (c_{0,n}u^0 + c_{1,n}u + c_{2,n}u^2 + c_{3,n}u^3 + \cdots + c_{q,n}u^q)x^n + \cdots.$$

Since each of the $c_{t,n}$ partitions that are counted by $c_{t,n}u^t x^n$ contains t k's, these partitions contribute $t \cdot c_{t,n}$ to the total number $T(k,n)$ of parts equal to k. Hence

$$T(k,n) = 0 \cdot c_{0,n} + 1 \cdot c_{1,n} + 2 \cdot c_{2,n} + 3 \cdot c_{3,n} + \cdots + q \cdot c_{q,n}.$$

Now, the term in x^n that is obtained by taking the partial derivative of $F(u,x)$ with respect to u is

$$(1 \cdot c_{1,n} + 2 \cdot c_{2,n}u + 3 \cdot c_{3,n}u^2 + \cdots + q \cdot c_{q,n}u^{q-1})x^n,$$

and with $u = 1$, the coefficient reduces to $T(k,n)$ itself. Therefore, denoting the coefficient of x^i in $g(x)$ with square brackets, $[x^i]g(x)$, we have simply that

$$T(k,n) = [x^n] \left. \frac{\partial F(u,x)}{\partial u} \right|_{u=1}.$$

It remains only to go through the straightforward calculations.

First, observe that the basic series

$$\begin{aligned} f(u,x) &= x + x^2 + \cdots + ux^k + \cdots \\ &= (u-1)x^k + (x + x^2 + \cdots + x^k + \cdots) \\ &= (u-1)x^k + x(1 + x + x^2 + \cdots) \\ &= (u-1)x^k + \frac{x}{1-x}. \end{aligned}$$

Hence

$$\begin{aligned} F(u,x) &= 1 + f(u,x) + [f(u,x)]^2 + [f(u,x)]^3 + \cdots \\ &= [1 - f(u,x)]^{-1} \\ &= \left[1 - (u-1)x^k - \frac{x}{1-x}\right]^{-1}, \end{aligned}$$

giving

$$\begin{aligned} \left.\frac{\partial F(u,x)}{\partial u}\right|_{u=1} &= (-1)\left[1 - (u-1)x^k - \frac{x}{1-x}\right]^{-2} \cdot (-x^k)\Bigg|_{u=1} \\ &= x^k\left(1 - \frac{x}{1-x}\right)^{-2} \\ &= x^k\left(\frac{1-2x}{1-x}\right)^{-2} \\ &= x^k(1-x)^2(1-2x)^{-2}. \end{aligned}$$

Finally, then,

$$\begin{aligned} T(k,n) &= [x^n]\left.\frac{\partial F(u,x)}{\partial u}\right|_{u=1} \\ &= [x^n]x^k(1-x)^2(1-2x)^{-2} \\ &= [x^{n-k}](1-2x+x^2)[1 + 2(2x) + 3(2x)^2 + \cdots \\ &\qquad + (i+1)(2x)^i + \cdots] \\ &= [x^{n-k}](1-2x+x^2)[\cdots + (n-k-1)2^{n-k-2}x^{n-k-2} \\ &\qquad + (n-k)2^{n-k-1}x^{n-k-1} + (n-k+1)2^{n-k}x^{n-k} + \cdots] \\ &= 1\cdot(n-k+1)2^{n-k} - 2\cdot(n-k)2^{n-k-1} + 1\cdot(n-k-1)2^{n-k-2} \\ &= (n-k+1)2^{n-k} - (n-k)2^{n-k} + (n-k-1)2^{n-k-2} \\ &= 2^{n-k} + (n-k-1)2^{n-k-2}. \end{aligned}$$

Hence

Finally, then

SECTION 19

Four Problems from the 1995 International Olympiad

1. In Figure 1, $ABCDEF$ is a convex hexagon in which

$$AB = BC = CD,$$
$$DE = EF = FA,$$

and

$$\angle BCD = \angle EFA = 60°.$$

G and H are points inside the hexagon such that AB subtends an angle of 120° at G and DE subtends an angle of 120° at H. Prove that

$$AG + GB + GH + HD + HE \geq FC.$$

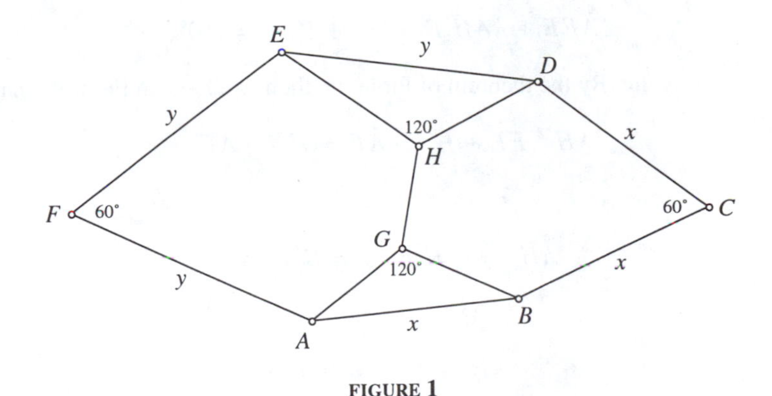

FIGURE 1

Since each of the triangles BCD and EFA is isosceles with a 60° angle between its equal sides, they are both equilateral, and in Figure 2 we have

$$BD = x = AB \quad \text{and} \quad AE = y = ED.$$

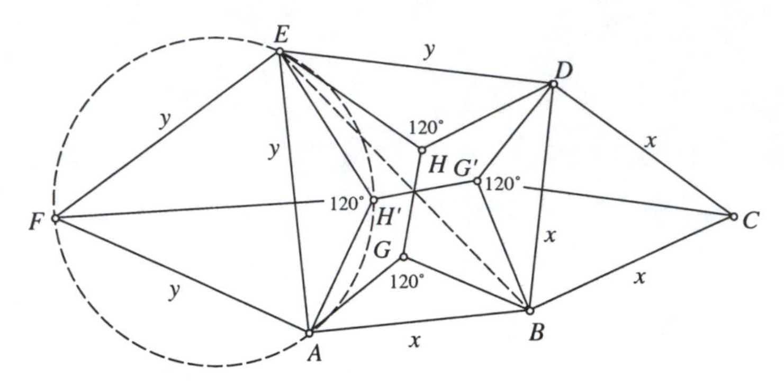

FIGURE 2

Thus B is equidistant from A and D, and so is E, making BE the perpendicular bisector of AD. Therefore, reflecting in BE interchanges A and D and takes G and H to images G' and H', preserving the lengths of the five segments in question in their respective images (Figure 2):

$$AG = G'D, \quad GB = G'B, \quad GH = G'H', \quad HD = H'A, \quad HE = H'E;$$

also

$$\angle AH'E = 120° = \angle BG'D.$$

Since

$$\angle AFE + \angle AH'E = 60° + 120° = 180°,$$

$AH'EF$ is cyclic. By the theorem of Ptolemy, then, we have in Figure 2 that

$$AH' \cdot FE + H'E \cdot AF = H'F \cdot AE,$$

that is

$$AH' \cdot y + H'E \cdot y = H'F \cdot y,$$

giving

$$AH' + H'E = H'F.$$

Similarly,

$$BG' + G'D = G'C.$$

Therefore,

$$\begin{aligned} AG + GB + GH + HD + HE \\ &= (DG' + G'B) + G'H' + (H'A + H'E) \\ &= G'C + G'H' + H'F \\ &= \text{polygonal path } FH'G'C \geq FC, \text{ as desired.} \end{aligned}$$

2. A, B, C, and D are distinct points on a straight line, in that order (Figure 3). The circles with diameters AC and BD intersect at X and Y. O is any point on the line XY except the point where it crosses AD. CO intersects the circle on diameter AC again at M, and BO intersects the circle on diameter BD again at N. Prove that AM, DN, and XY are concurrent.

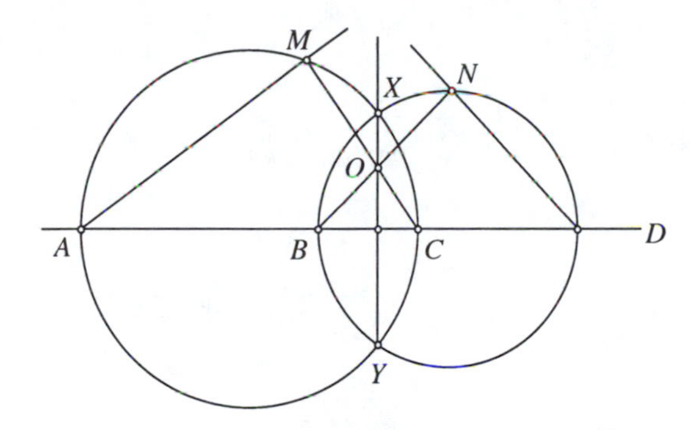

FIGURE 3

Suppose XY meets AD at K, AM at P, and DN at Q (Figure 4). We would like to show that P and Q are the same point, which would be the case if we could show that $KP = KQ$.

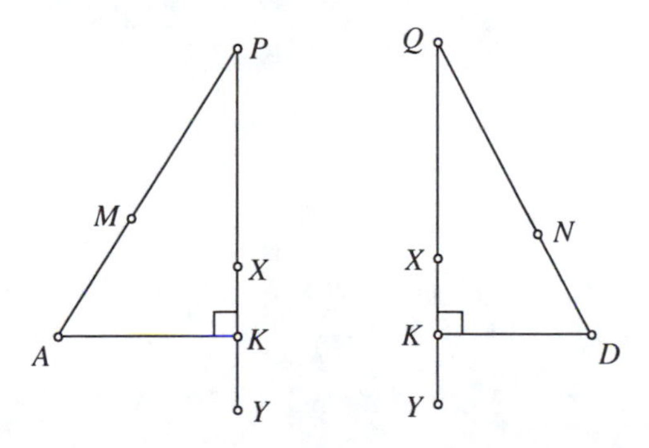

FIGURE 4

Since XY is perpendicular to AD, we have

$$PK = AK \cdot \tan \angle A \quad \text{and} \quad QK = DK \cdot \tan \angle D;$$

thus we would like to show that

$$AK \cdot \tan \angle A = DK \cdot \tan \angle D. \tag{1}$$

Now, the right angles at M and K imply that $MAKO$ is cyclic (Figure 5), giving

$$\text{interior } \angle A = \text{exterior } \angle KOC;$$

similarly, $NOKD$ is cyclic and

$$\angle D = \angle KOB.$$

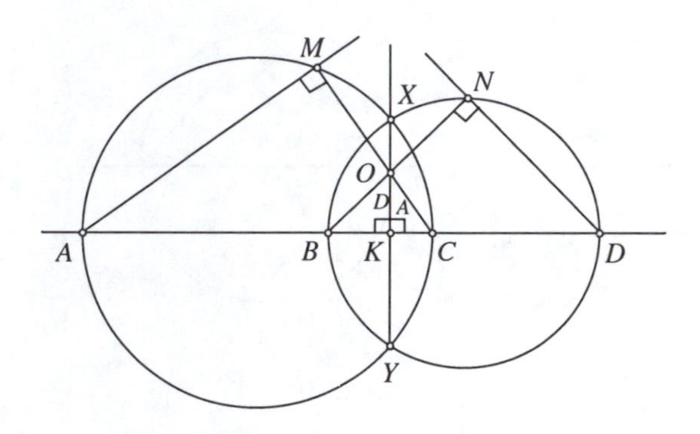

FIGURE 5

From right triangles KOC and KOB, then, condition (1) amounts to showing that

$$AK \cdot \frac{KC}{KO} = KD \cdot \frac{BK}{KO},$$

that is,

$$AK \cdot KC = KD \cdot BK.$$

Now, chords AC and XY intersect at K in the one circle and chords BD and XY intersect at K in the other. Thus

$$AK \cdot KC = XK \cdot KY = BK \cdot KD,$$

and the conclusion follows.

(We observe that the problem is equivalent to showing that $MADN$ is cyclic.)

3. Determine all integers $n > 3$ for which there exists in the plane a set of n points $A_1, A_2, \ldots, A_n$, no three collinear, and a corresponding set of n real numbers $p_1, p_2, \ldots, p_n$ such that the area of

$$\triangle A_i A_j A_k = p_i + p_j + p_k \quad \text{for all } 1 \leq i < j < k \leq n.$$

It is easy to see that $n = 4$ is acceptable, for the area of the triangle determined by each three vertices of a parallelogram $ABCD$ is $\frac{1}{2}ABCD$, and hence, setting

$$p_i = \frac{1}{6}ABCD \quad \text{for all } i = 1, 2, 3, 4,$$

completes a satisfactory solution. Thus, let $n \geq 5$.

Since no three of the points are collinear, each subset of four of the points must lie in the plane so that either (i) one of the points lies inside the triangle formed by the other three, or (ii) the four points determine a convex quadrilateral.

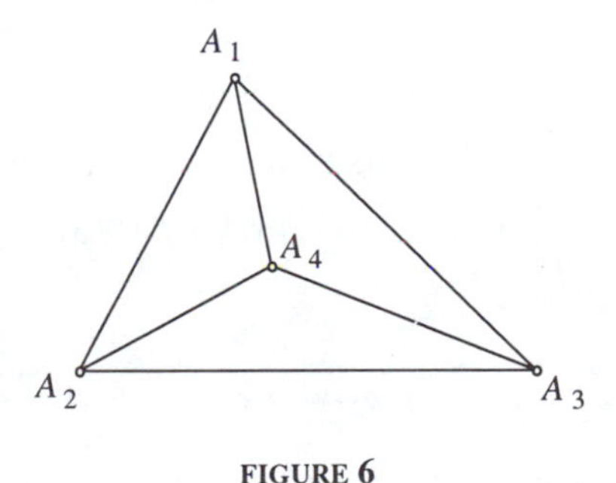

FIGURE 6

Case (i): Suppose that A_4 lies inside $\triangle A_1A_2A_3$ (Figure 6). Since $\triangle A_1A_2A_3$ is decomposed by segments to the vertices from A_4, we have

$$\triangle A_1A_2A_4 + \triangle A_1A_3A_4 + \triangle A_2A_3A_4 = \triangle A_1A_2A_3,$$
$$(p_1 + p_2 + p_4) + (p_1 + p_3 + p_4) + (p_2 + p_3 + p_4) = (p_1 + p_2 + p_3),$$

giving

$$p_4 = -\frac{1}{3}(p_1 + p_2 + p_3) = -\frac{1}{3}\triangle A_1A_2A_3.$$

That is to say, if A_t is inside a triangle $A_iA_jA_k$, the value of p_t is negative and of magnitude equal to $\frac{1}{3}$ the area of the triangle. Thus each triangle $A_iA_jA_k$ that contains A_t in its interior has the **same** area.

Now we ask where a fifth point A_5 might occur in the plane. Since no three points A_i can be collinear, A_5 cannot avoid determining a triangle, with some two of the three points A_1, A_2, A_3, that contains A_4 in its interior (Figure 7); suppose $\triangle A_5A_2A_3$ contains A_4.

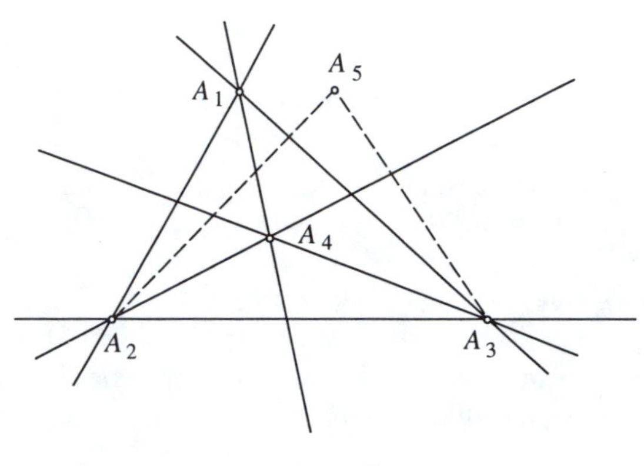

FIGURE 7

This makes

$$\triangle A_5A_2A_3 = \triangle A_1A_2A_3,$$
$$p_5 + p_2 + p_3 = p_1 + p_2 + p_3,$$

and

$$p_5 = p_1.$$

Also, A_5 determines the convex quadrilaterals $A_5A_1A_2A_4$ and $A_5A_1A_4A_3$. Now, each diagonal decomposes a convex quadrilateral, and for $A_5A_1A_2A_4$ we have (Figure 8)

$$\triangle A_1A_2A_5 + \triangle A_5A_2A_4 = \triangle A_1A_2A_4 + \triangle A_1A_4A_5,$$
$$(p_1 + p_2 + p_5) + (p_5 + p_2 + p_4) = (p_1 + p_2 + p_4) + (p_1 + p_4 + p_5),$$

and

$$p_5 + p_2 = p_1 + p_4.$$

That is to say,

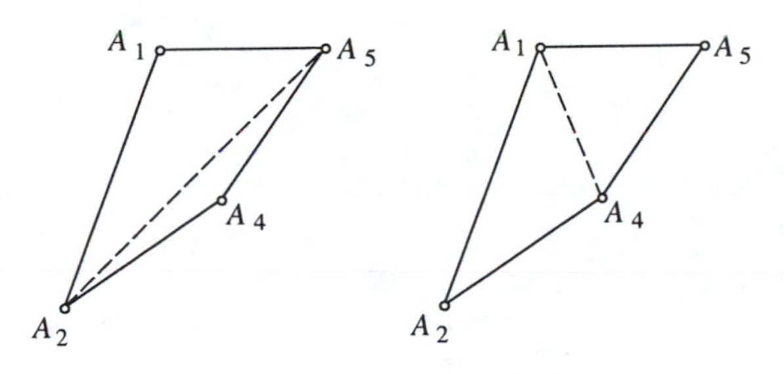

FIGURE 8

for any convex quadrilateral, the sum of the p-values of each pair of opposite vertices is the same.

Since $p_5 = p_1$, this equation gives

$$p_2 = p_4.$$

Similarly, from $A_5 A_1 A_4 A_3$ we get

$$p_3 = p_4.$$

Hence

$$p_2 = p_3 = p_4,$$

and since p_4 is negative, this gives an impossible negative area for $\triangle A_2 A_3 A_4$. Thus we conclude that if any point A_i lies inside the triangle formed by three others, then n cannot exceed 4.

Case (ii): Suppose, then, that no A_i lies inside a triangle formed by three others. In this case, every quadrilateral $A_i A_j A_k A_t$ must be convex (case (ii) above), and for the pairs of opposite vertices,

$$p_i + p_k = p_j + p_t.$$

Finally, consider any subset of five of the points, say A_1, A_2, A_3, A_4, A_5. From the convex quadrilaterals $A_1 A_2 A_3 A_4$ and $A_1 A_2 A_3 A_5$, we have

$$p_1 + p_3 = p_2 + p_4 \quad \text{and} \quad p_1 + p_3 = p_2 + p_5.$$

Hence

$$p_2 + p_4 = p_2 + p_5,$$

giving

$$p_4 = p_5.$$

But p_4 and p_5 are not special; we similarly obtain $p_i = p_j$ for all pairs, and it follows that all five p_i have the same value. As a result, all the triangles determined by these five A_i have the same area.

Now consider the convex hull H of these five A_i. Suppose A_1A_2 is a side of H. In this case, A_3, A_4, A_5 all lie on the same side of A_1A_2 (Figure 9), and the equality of the three triangles on base A_1A_2, namely $\triangle A_1A_2A_3$, $\triangle A_1A_2A_4$, and $\triangle A_1A_2A_5$, implies that the third vertices are collinear on a line which is parallel to A_1A_2, a contradiction. Thus, also in this case, the set $A_1, A_2, \ldots, A_n$ cannot contain as many as five points, and we conclude that $n = 4$ is the only value of n.

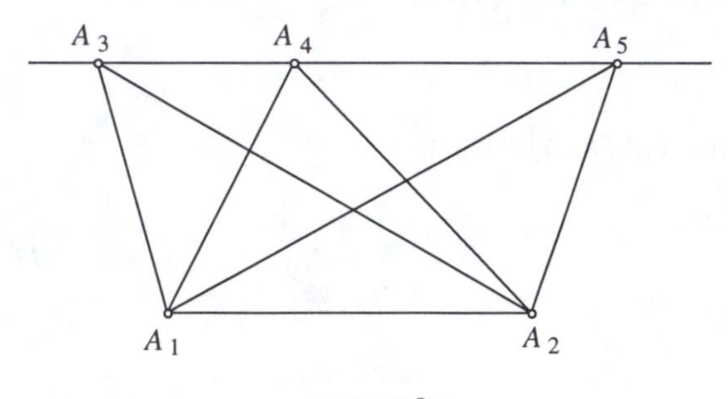

FIGURE 9

Finally, here is a beautiful solution due to my good friend and colleague Ian Goulden.

4. If p is an odd prime, how many p-element subsets of $\{1, 2, 3, \ldots, 2p\}$ are there such that the sum of all the elements in the subset is divisible by p?

Consider the generating function

$$\begin{aligned} F(u, x) &= (1 + ux)(1 + ux^2)(1 + ux^3) \cdots (1 + ux^{2p}) \\ &= \cdots + a_{m,n}u^m x^n + \cdots . \end{aligned}$$

In multiplying out, from each factor $(1+ux^k)$ one takes either the 1 or the ux^k, multiplies the $2p$ selections and adds the results of doing this in all possible ways. After some simplification, the coefficient $a_{m,n}$ of the term $a_{m,n}u^m x^n$ is the number of ways of selecting an m-element subset from the slate of exponents $\{1, 2, 3, \ldots, 2p\}$ so that the sum of the m elements is n. Hence our task is to evaluate the sum

$$S = \sum_{p|n} a_{p,n} = a_{p,0} + a_{p,p} + a_{p,2p} + a_{p,3p} + \cdots.$$

Now, there are generally various values of m which contribute to the construction of terms in x^n; for example, from $(1+ux)(1+ux^2)\cdots(1+ux^{10})$ a term in x^7 is obtained alternatively as

$$1^9 \cdot ux^7, \quad 1^8 \cdot ux \cdot ux^6 = u^2x^7, \quad \text{and} \quad 1^7 \cdot ux \cdot ux^2 \cdot ux^4 = u^3x^7,$$

and in various other ways. Thus the total coefficient of x^n is a function of u and we may write

$$F(u, x) = F(x) = \sum_{n \geq 0} f_n(u)x^n.$$

The required sum S is the sum of the coefficients of the terms in u^p in the functions $f_n(u)$ for the values of n which are divisible by p. Using square brackets to denote the coefficient of y^k in the function $g(y)$, as in $[y^k]g(y)$, then

$$S = \sum_{p|n} [u^p] f_n(u).$$

Since it doesn't matter whether we work out these coefficients individually and add them up or first add the functions and then take the coefficient of u^p in the sum, we have

$$S = [u^p] \sum_{p|n} f_n(u).$$

Clearly, the terms which contribute to the sum S constitute every pth term in the series

$$F(x) = f_0(u)x^0 + f_1(u)x + f_2(u)x^2 + \cdots.$$

Now, there is a nice formula for the sum of every nth term of a series. Since this formula is proved in my *Mathematical Gems III* (pages 210–214), let us simply put it to good use here. The result is the following:

if

$$F(x) = f_0 + f_1x + f_2x^2 + \cdots + f_{n-1}x^{n-1} + f_nx^n + \cdots$$

is a finite or infinite series, the sum $S(n, j)$ of every nth term, beginning at $f_j x^j$, where $j \le n - 1$, that is,

$$S(n, j) = f_j x^j + f_{j+n} x^{j+n} + f_{j+2n} x^{j+2n} + \cdots,$$

is given by the formula

$$S(n, j) = \frac{1}{n} \sum_{t=0}^{n-1} \omega^{-jt} f(\omega^t x),$$

where $\omega =$ the complex nth root of unity $e^{2\pi i/n}$.

(Comment: There is a misleading inaccuracy in the discussion on page 211 of *Mathematical Gems III*. The first two lines should say that $S(n, j)$ begins at the $(j + 1)$th term and that $j \le n - 1$.)

In the case at hand we have $n = p$ and $j = 0$, making $\omega^{-jt} = 1$ and giving

$$\sum_{p|n} f_n(u) x^n = \frac{1}{p} \sum_{t=0}^{p-1} F(\omega^t x).$$

Putting $x = 1$ does not effect the coefficients and we have

$$S = [u^p] \sum_{p|n} f_n(u) = [u^p] \frac{1}{p} \sum_{t=0}^{p-1} F(\omega^t).$$

Recalling that

$$F(x) = (1 + ux)(1 + ux^2) \cdots (1 + ux^{2p}),$$

we have

$$F(\omega^t) = (1 + u\omega^t)(1 + u(\omega^t)^2)(1 + u(\omega^t)^3) \cdots (1 + u(\omega^t)^{2p}),$$

and it is at this point that the prime character of p comes into play. Since p is an odd prime, it follows that, except for $t = 0$, the powers

$$\omega^t, (\omega^t)^2, (\omega^t)^3, \ldots, (\omega^t)^p$$

constitute, in some order, the p pth roots of unity $1, \omega, \omega^2, \ldots, \omega^{p-1}$; similarly the powers $(\omega^t)^{p+1}, (\omega^t)^{p+2}, \ldots, (\omega^t)^{2p}$ yield the same set of values. Thus, for all $t = 1, 2, \ldots, p - 1$,

$$F(\omega^t) = [(1 + u)(1 + u\omega)(1 + u\omega^2) \ldots (1 + u\omega^{p-1})]^2.$$

Now let us show that this product is just $(1 + u^p)^2$.

It is easy to see that the roots of the equation

$$1 - x^p = 0 \quad \text{are} \quad 1, \omega, \omega^2, \ldots, \omega^{p-1},$$

yielding the factoring

$$(1-x)(1-\omega x)(1-\omega^2 x)\cdots(1-\omega^{p-1}x) = 1 - x^p:$$

substituting $x = \omega^{p-k}$ makes the factor $(1 - \omega^k x)$ on the left side equal to $1 - \omega^p = 1 - 1 = 0$ and makes the right side $1 - x^p = 1 - (\omega^p)^{k-p} = 1 - 1 = 0$. For $x = -u$, then, we obtain

$$(1+u)(1+u\omega)(1+u\omega^2)\cdots(1+u\omega^{p-1}) = 1 - (-u)^p = 1 + u^p,$$

since p is odd. For each of the $p-1$ values $t = 1, 2, \ldots, p-1$, then, we get the same result

$$F(\omega^t) = (1+u^p)^2 = 1 + 2u^p + u^{2p}.$$

Finally, for $t = 0$, we have simply

$$F(\omega^t) = F(1) = (1+u)(1+u)\cdots(1+u) = (1+u)^{2p},$$

and altogether, then,

$$\begin{aligned}
S &= [u^p]\frac{1}{p}\sum_{t=0}^{p-1} F(\omega^t) \\
&= \frac{1}{p}[u^p]\{(1+u)^{2p} + (p-1)(1+2u^p+u^{2p})\} \\
&= \frac{1}{p}\left\{\binom{2p}{p} + (p-1)\cdot 2\right\}.
\end{aligned}$$

SECTION 20

Two Geometry Problems

1. In $\triangle ABC$, $\angle B$ is twice $\angle C$, and P is the point inside the triangle where the perpendicular bisector of BC is intersected by the circle with center A and radius AB (Figure 1). Prove that AP **trisects** $\angle A$.

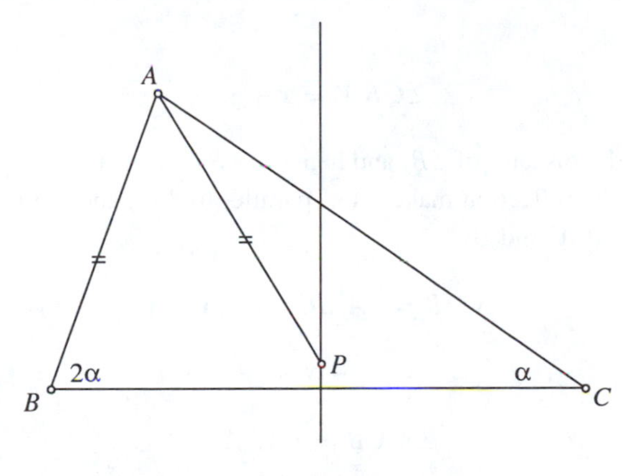

FIGURE 1

The following beautiful solution is due to Adam Brown, a third year student at the University of Waterloo (in 1995).

Since P is on the perpendicular bisector of BC, $\triangle PBC$ is isosceles. Let the equal base angles in this triangle be x and let $\angle ACP = y$ (Figure 2). Then, in $\triangle ABC$,

$$\angle C = x + y \quad \text{and} \quad \angle B = 2\angle C = 2x + 2y.$$

Now let the figure be reflected in the perpendicular bisector of BC. This carries AP into $A'P$, and so

$$AP = A'P.$$

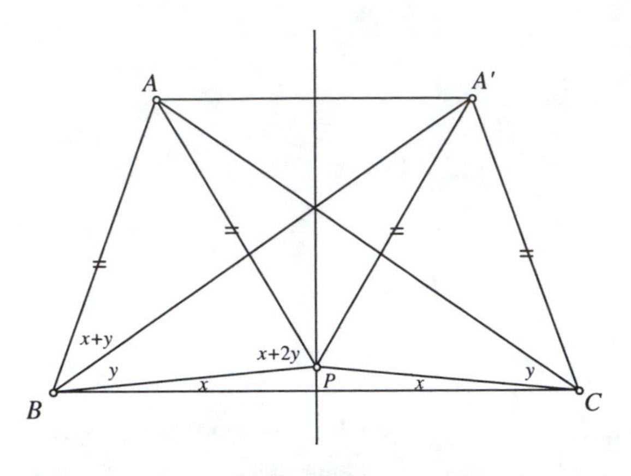

FIGURE 2

Also, since this interchanges B and C, angle C is carried to angle CBA', and we have

$$\angle CBA' = x + y,$$

making BA' the bisector of $\angle B$, and hence $\angle ABA' = x + y$.

Clearly, the reflection makes AA' parallel to BC, and we have equal alternate angles at A' and B:

$$\angle AA'B = \angle A'BC \quad (= x + y).$$

Therefore

$$\angle AA'B = \angle ABA',$$

and $\triangle ABA'$ is isosceles with $AB = AA'$. Thus $\triangle APA'$ is actually equilateral and $\angle APA' = 60°$.

The base angles in isosceles $\triangle ABP$ are $x + 2y$, and since $\triangle ABP$ is congruent to its image $\triangle A'PC$, then $\angle A'PC$ is also $x + 2y$. The sum of the angles around P, then, is

$$\begin{aligned} 360° &= \angle APB + \angle BPC + \angle A'PC + \angle APA' \\ &= (x + 2y) + (180° - 2x) + (x + 2y) + 60° \\ &= 240° + 4y, \end{aligned}$$

revealing the remarkable result that, for all configurations of this kind, the angle y is always $30°$.

Finally, then, in $\triangle ABC$,

$$\angle A = 180^\circ - (3x + 3y) = 90^\circ - 3x,$$

and in $\triangle APC$,

$$\begin{aligned}\angle PAC &= 180^\circ - y - 60^\circ - (x + 2y)\\ &= 30^\circ - x \qquad (\text{since } y = 30^\circ)\\ &= \frac{1}{3}\angle A.\end{aligned}$$

2. This problem comes to us through the courtesy of Daniel Herling, a first-year student at the University of Waterloo (in 1995).

   Triangle XYZ is inscribed in $\triangle ABC$ with X on BC, Y on CA, and Z on AB so that each angle is the same as the angle at the opposite vertex:

$$\angle X = \angle A,\ \angle Y = \angle B, \text{ and } \angle Z = \angle C \quad \text{(Figure 3).}$$

   (a) Prove that

   the circumcenter O of $\triangle ABC$ is the orthocenter of $\triangle XYZ$.

   (b) Of all such triangles XYZ, prove the one having smallest perimeter is the medial triangle of $\triangle ABC$, i.e., with vertices at the midpoints of the sides of $\triangle ABC$.

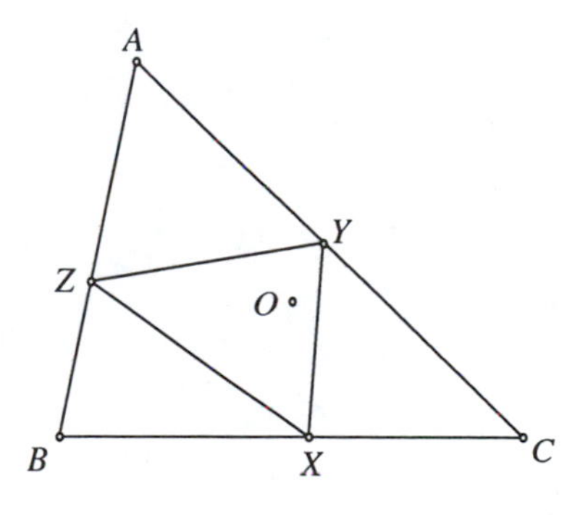

FIGURE 3

(a) By a theorem of Miquel, for any points X, Y, Z, one on each side of $\triangle ABC$, the circumcircles of triangles AZY, BXZ, and CYX are concurrent at a point called the Miquel point of $\triangle XYZ$. The proof of this theorem is immediate:

Suppose the circles around triangles BXZ and CYX meet again at P (Figure 4). Since the exterior angle at a vertex of a cyclic quadrilateral

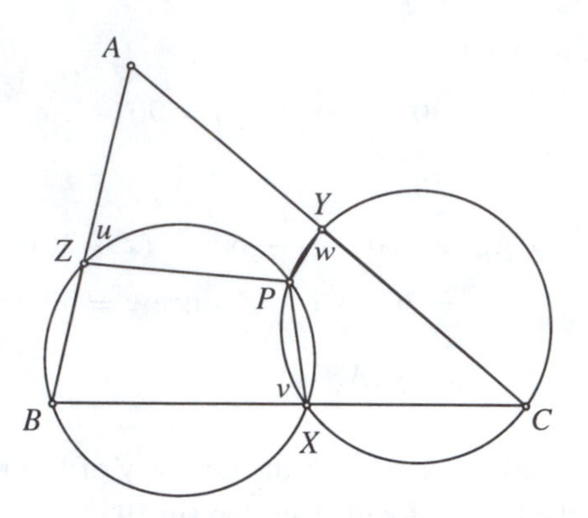

FIGURE 4

is equal to the interior angle at the opposite vertex, we have $u = v$ and $v = w$, and so $u = w$, implying that $AZPY$ is also cyclic.

Now, the Miquel point enjoys an interesting property concerning the angles which are subtended at it by the sides of the parent triangle:

the angle subtended at P by a side of $\triangle ABC$ is equal to the sum of the angle in $\triangle ABC$ which is opposite that side and the angle between the two sides of $\triangle XYZ$ which meet on that side; e.g., for BC in Figure 5,

$$S = A + t.$$

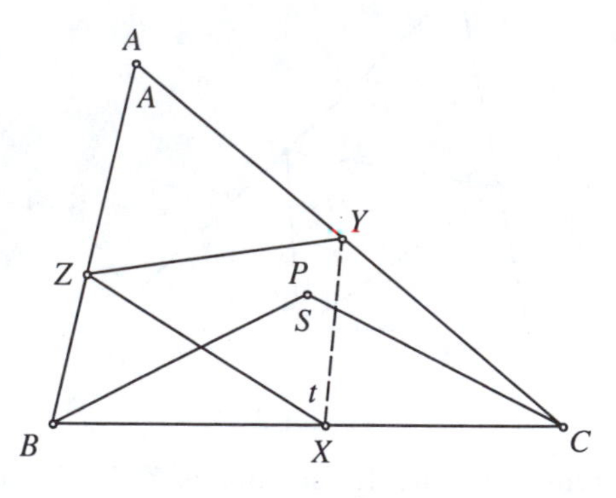

FIGURE 5

This is easily proved as follows.

In Figure 6, the exterior angles q and r are equal to $d + e$ and $f + g$ respectively, giving

$$S = q + r = d + e + f + g = A + e + g.$$

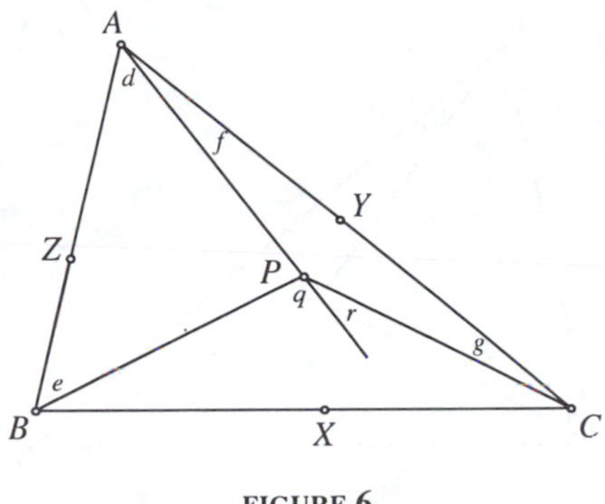

FIGURE 6

But, in cyclic quadrilateral $ZBXP$ (Figure 7), ZP subtends equal angles e at B and X; similarly PY subtends equal angles g at X and C, and we have $e + g = t$, making

$$S = A + t, \text{ as claimed.}$$

Now, in the problem at hand, the angle t at X in $\triangle XYZ$ is also $\angle A$, and so $\angle BPC = 2\angle A$. Therefore the point P lies on the arc of a circle drawn on BC so as to contain the angle $2\angle A$. Similarly, P lies on the arc of a circle drawn on CA so as to contain the angle $2\angle B$. But, since each side of a triangle subtends at the circumcenter of the triangle an angle that is twice the opposite angle in the triangle, the circumcenter O of $\triangle ABC$ also lies on each of these arcs. Since these arcs meet in only one other point in addition to C, it follows that the circumcenter O of $\triangle ABC$ is in fact the Miquel point P of $\triangle XYZ$.

Now that we know P is the circumcenter of $\triangle ABC$, it is easy to show that it is also the orthocenter of $\triangle XYZ$. In Figure 8, each of AP, BP, CP

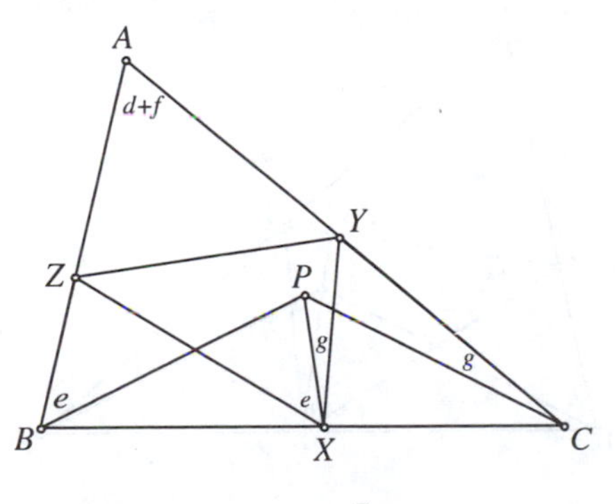

FIGURE 7

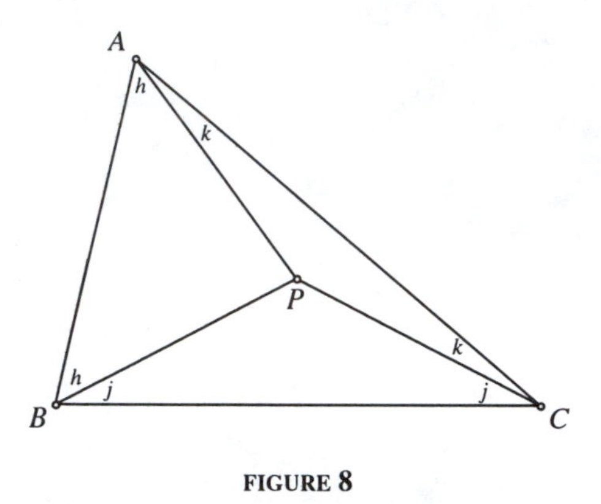

FIGURE 8

is the circumradius of $\triangle ABC$, making triangles APB, BPC, CPA isosceles, with pairs of equal base angles h, j, and k respectively.

Since $2h + 2j + 2k = 180°$, then $h + j + k = 90°$.

In Figure 9, let XP meet ZY at Q. Now,

$$\angle Z = \angle C = j + k, \quad \text{and} \quad \angle ZBP = h = \angle ZXP$$

(in cyclic quadrilateral $ZBXP$),

and therefore, in $\triangle XZQ$, the angles at X and Z add up to $h + j + k = 90°$, making the angle at Q a right angle and placing P on the altitude of $\triangle XYZ$ from X. Similarly, P lies on the altitudes from Y and Z and is therefore the orthocenter of $\triangle XYZ$.

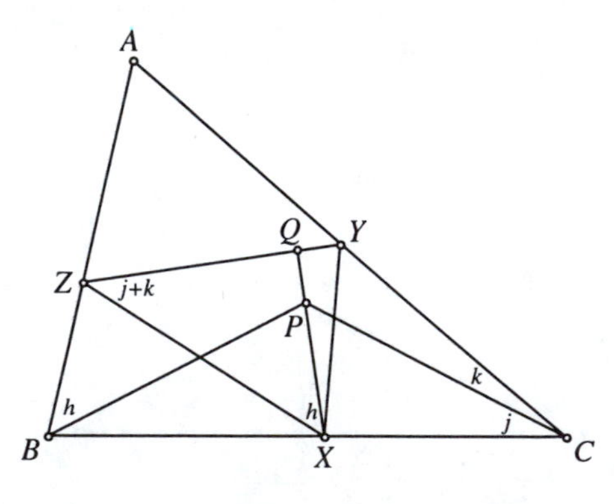

FIGURE 9

(b) Since the angles of $\triangle XYZ$ are the same as those of $\triangle ABC$, all such triangles XYZ are similar to $\triangle ABC$, and the length of XP is a measure of the size of $\triangle XYZ$:

> Imagine all the triangles XYZ piled on top of one another with their orthocenters at the same point P and their respective sides parallel (Figure 10). Clearly the smallest triangle has the shortest PX.

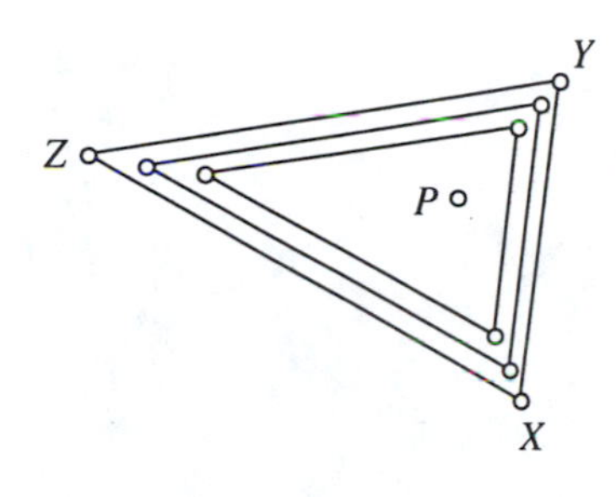

FIGURE 10

Thus the $\triangle XYZ$ of smallest perimeter (and area) occurs when X is the foot of the perpendicular from P to BC. Now, the perpendicular from the circumcenter to a side bisects the side, in which case X would be the midpoint of BC; similarly for Y and Z, making $\triangle XYZ$ the medial triangle. Therefore the smallest $\triangle XYZ$ would indeed be the medial triangle of $\triangle ABC$ if it actually belongs to the family of triangles XYZ. Unfortunately, it is conceivable that there might not be any $\triangle XYZ$ with X at the midpoint of BC. However, it is easy to see that the medial triangle **does** belong to this family, for the sides of the medial triangle are parallel to the respective sides of the parent triangle, making each angle of the medial triangle equal to the opposite angle in $\triangle ABC$.

SECTION 21

An Unlikely Perfect Square

Prove that

$$S_n = \sum_{k=1}^{2n-1} 2^{k-1} \binom{4n-2}{2k}$$
$$= 2^0 \binom{4n-2}{2} + 2\binom{4n-2}{4} + 2^2 \binom{4n-2}{6} + \cdots$$
$$+ 2^{2n-2} \binom{4n-2}{4n-2}$$

is a perfect square for all $n = 1, 2, 3, \ldots$.

For example,

$$S_2 = 1\binom{6}{2} + 2\binom{6}{4} + 4\binom{6}{6} = 15 + 30 + 4 = 49.$$

This is a problem of the ingenious Leo Moser (University of Alberta) which appeared in *Mathematics Magazine*, 1964, 61; the clever solution is due to Henry W. Gould (University of West Virginia).

Adding the binomial expansions

$$(1+\sqrt{x})^{2t} = 1 + \binom{2t}{1}\sqrt{x} + \binom{2t}{2}x + \binom{2t}{3}x\sqrt{x} + \cdots + \binom{2t}{2t}x^t,$$
$$(1-\sqrt{x})^{2t} = 1 - \binom{2t}{1}\sqrt{x} + \binom{2t}{2}x - \binom{2t}{3}x\sqrt{x} + \cdots + \binom{2t}{2t}x^t,$$

eliminates the terms in $\sqrt{x}$, giving

$$(1+\sqrt{x})^{2t} + (1-\sqrt{x})^{2t} = 2\left[1 + \binom{2t}{2}x + \binom{2t}{4}x^2 + \cdots + \binom{2t}{2t}x^t\right].$$

Now

$$S_n = \binom{4n-2}{2}2^0 + \binom{4n-2}{4}2 + \binom{4n-2}{6}2^2 + \cdots + \binom{4n-2}{4n-2}2^{2n-2},$$

and hence, for $2t = 4n - 2$ and $x = 2$, we have

$$\begin{aligned}(1+\sqrt{2})^{4n-2} &+ (1-\sqrt{2})^{4n-2} \\ &= 2\left[1+\binom{4n-2}{2}2+\binom{4n-2}{4}2^2+\cdots+\binom{4n-2}{4n-2}2^{2n-1}\right] \\ &= 2(1+2S_n),\end{aligned}$$

from which

$$\begin{aligned} S_n &= \frac{(1+\sqrt{2})^{4n-2}+(1-\sqrt{2})^{4n-2}-2}{4} \\ &= \left[\frac{(1+\sqrt{2})^{2n-1}+(1-\sqrt{2})^{2n-1}}{2}\right]^2 \end{aligned}$$

(note $(1+\sqrt{2})(1-\sqrt{2}) = -1$).

But, just as the terms in $\sqrt{x}$ cancel in the sum $(1+\sqrt{x})^m + (1-\sqrt{x})^m$, the irrational terms cancel in $(1+\sqrt{2})^{2n-1} + (1-\sqrt{2})^{2n-1}$ and the others double up, implying that

$$Sn = \left[\frac{(1+\sqrt{2})^{2n-1}+(1-\sqrt{2})^{2n-1}}{2}\right]^2$$

is indeed the square of an integer.

$$\left(\text{Thus } S_2 = \left[\frac{(1+\sqrt{2})^3+(1-\sqrt{2})^3}{2}\right]^2 = (1+3\cdot 2)^2 = 7^2 = 49.\right)$$

SECTION 22

The Nine-Point Circle and Coolidge's Theorem, the De Longchamps Point of a Triangle, Cantor's Theorem, and Napoleon's Theorem

Except for the discussion of the De Longchamps Point (noted in [1]), this essay owes its inspiration to Liang-shin Hahn's outstanding book *Complex Numbers and Geometry* (MAA, Spectrum Series, 1994).

1. The Nine-Point Circle and Coolidge's Theorem

The midpoints of the segments joining the orthocenter of a triangle to the vertices are called the **Euler** points of the triangle (in Figure 1, H is the orthocenter and X, Y, and Z are the Euler points). Remarkably, then,

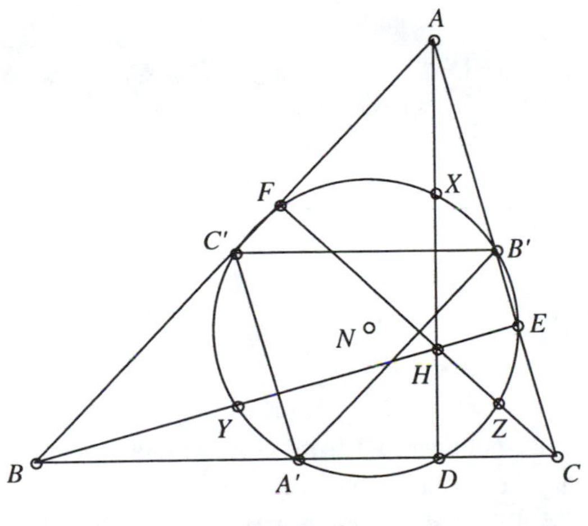

FIGURE 1

in any triangle, the nine points consisting of the three midpoints of the sides, the three feet of the altitudes and the three Euler points, lie on a circle.

Proof: Let the circumcenter O of $\triangle ABC$ be the origin of vectors and let the vectors to the vertices A, B, and C, be α, β, and γ respectively (Figure 2).

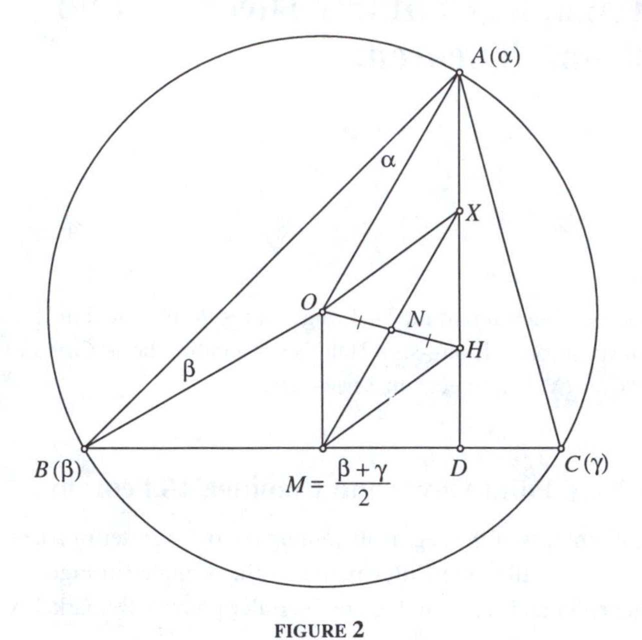

FIGURE 2

Now we ask "Where is the point $S = \alpha + \beta + \gamma$?

Clearly the vector **OM** from the center of the circumcircle to the midpoint M of the chord BC is perpendicular to BC. Now, $\alpha + \beta + \gamma$ may be written in the form

$$\alpha + 2\left(\frac{\beta + \gamma}{2}\right),$$

and so

$$S = \alpha + 2\left(\frac{\beta + \gamma}{2}\right) = \alpha + 2(\mathbf{OM}).$$

That is to say, one would get to S from O by first proceeding along the vector α to vertex A and then going in the direction of **OM** a distance equal to twice the length of **OM**. But **OM** is **perpendicular** to BC, and therefore S must lie on the altitude from A. Similarly S also lies on the other two altitudes and we conclude that S must in fact be the **orthocenter** H of $\triangle ABC$.

Now, the Euler point X bisects AH (Figure 2), which is parallel to **OM** and twice as long, and so **OM** is equal and parallel to each of AX and XH, making $OMHX$ and $OMXA$ parallelograms. Suppose the diagonals of $OMHX$ bisect each other at N. Then

$$|NM| = |NX| = \frac{1}{2}|MX| = \frac{1}{2}|\alpha| \quad \text{(in parallelogram } OMXA) = \frac{1}{2}R,$$

where R is the circumradius of $\triangle ABC$.

Now, $\triangle MDX$ is right angled at D, and so the midpoint N of its hypotenuse (being its circumcenter) is equidistant from the three vertices.

Hence

$$|ND| = |NM| = |NX| = \frac{1}{2}R,$$

implying the circle with center N and radius $\frac{1}{2}R$ goes through D, M, and X.

Similarly, this circle goes through the other triads of feet, midpoints and Euler points, and the proof is complete.

We note that the radius of the nine-point circle is $\frac{1}{2}R$, and while its center N is conveniently located at the midpoint of OH, the above formulas for the coordinate vectors of H and N,

$$H = \alpha + \beta + \gamma \quad \text{and} \quad N = \frac{1}{2}(\alpha + \beta + \gamma),$$

are valid only when the circumcenter O is the origin of vectors and α, β, γ are the coordinate vectors of the vertices A, B, C.

Finally, we state explicitly the following useful result that also comes out of the above argument:

the distance along an altitude from a vertex to the orthocenter is twice the distance from the circumcenter to the opposite side (e.g., $AH = 2 \cdot OM$).

2. Coolidge's Theorem

(a) Let z_1, z_2, z_3, z_4, be any four points on a circle of radius R (Figure 3). They determine four triangles, each having the center O of the circle as circumcenter, and therefore, taking O as the origin of vectors, their nine-point centers N_1, N_2, N_3, N_4, are:

$$N_1 = \frac{1}{2}(z_2 + z_3 + z_4), \quad \text{for} \quad \Delta z_2 z_3 z_4,$$

$$N_2 = \frac{1}{2}(z_1 + z_3 + z_4), \quad \text{for} \quad \Delta z_1 z_3 z_4,$$

$$N_3 = \frac{1}{2}(z_1 + z_2 + z_4), \quad \text{for} \quad \Delta z_1 z_2 z_4,$$

and

$$N_4 = \frac{1}{2}(z_1 + z_2 + z_3), \quad \text{for} \quad \Delta z_1 z_2 z_3.$$

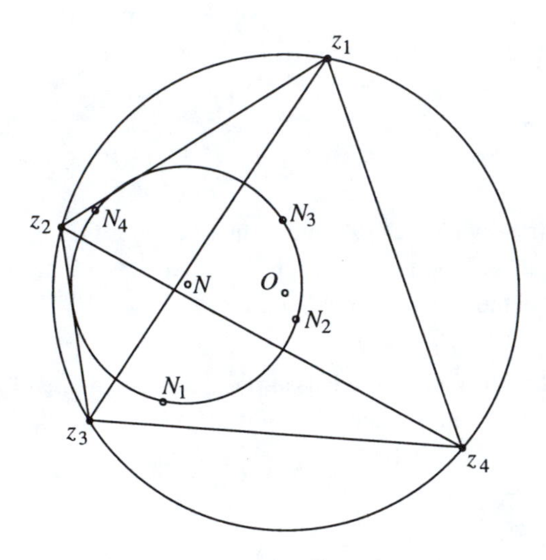

FIGURE 3

Now, the point

$$N = \frac{1}{2}(z_1 + z_2 + z_3 + z_4)$$
$$= N_1 + \frac{1}{2}z_1 = N_2 + \frac{1}{2}z_2 = N_3 + \frac{1}{2}z_3 = N_4 + \frac{1}{2}z_4.$$

Hence for all $i = 1, 2, 3, 4$, we have

$$|N - N_i| = \left|\frac{1}{2}z_i\right| = \frac{1}{2}R.$$

Thus the four nine-point centers N_i are themselves concyclic on a circle the same size as their own nine-point circles and having center N. N is called the nine-point center of the quadrilateral $z_1z_2z_3z_4$ and the circle through $N_1N_2N_3N_4$ is the nine-point circle of the quadrilateral $z_1z_2z_3z_4$.

Definition: Let $P = z_1 z_2, \ldots, z_n$ be any cyclic n-gon whose circumcenter is the origin of vectors and whose circumcircle has radius R. Then the point

$$N = \frac{1}{2}(z_1 + z_2 + \cdots + z_n)$$

is called the **nine-point center** of P and the circle with center N and radius $\frac{1}{2}R$ is the **nine-point circle** of P.

Coolidge's Theorem

> *For any cyclic n-gon $P = z_1 z_2 \ldots z_n$, the nine-point centers of the $(n-1)$-gons determined by suppressing the vertices of P in turn all lie on the nine-point circle of P.*

Proof: Let $S = z_1 + z_2 + \cdots + z_n$. Then the nine- point center of P is $N = \frac{1}{2}S$ and the nine-point center N_i of the $(n-1)$-gon determined by suppressing z_i is $N_i = \frac{1}{2}(S - z_i)$. Hence, for all i, we have

$$N = N_i + \frac{1}{2}z_i,$$

implying

$$|N - Ni| = \left|\frac{1}{2}z_i\right| = \frac{1}{2}|z_i|,$$

which places N_i on the circle with center N and half the radius of the circumcircle of P, which by definition is the nine-point circle of P.

We observe that the nine-point circles of the n-gons in a circle of radius R all have the same radius $\frac{1}{2}R$, for all $n = 3, 4, 5, \ldots$.

3. The De Longchamps Point

(i) Definition: The **de Longchamps point** L of a triangle is obtained by reflecting H in O (i.e. the orthocenter in the circumcenter; equivalently, give H a halfturn about O) (Figure 4).

(ii) Definition: The **anticomplementary triangle** of $\triangle ABC$ is the triangle XYZ whose medial triangle is $\triangle ABC$ (i.e., A, B, and C are the midpoints of the sides of $\triangle XYZ$).

We observe that $\triangle XYZ$ is determined by drawing through each of A, B, and C a line parallel to the opposite side of $\triangle ABC$ (Figure 4).

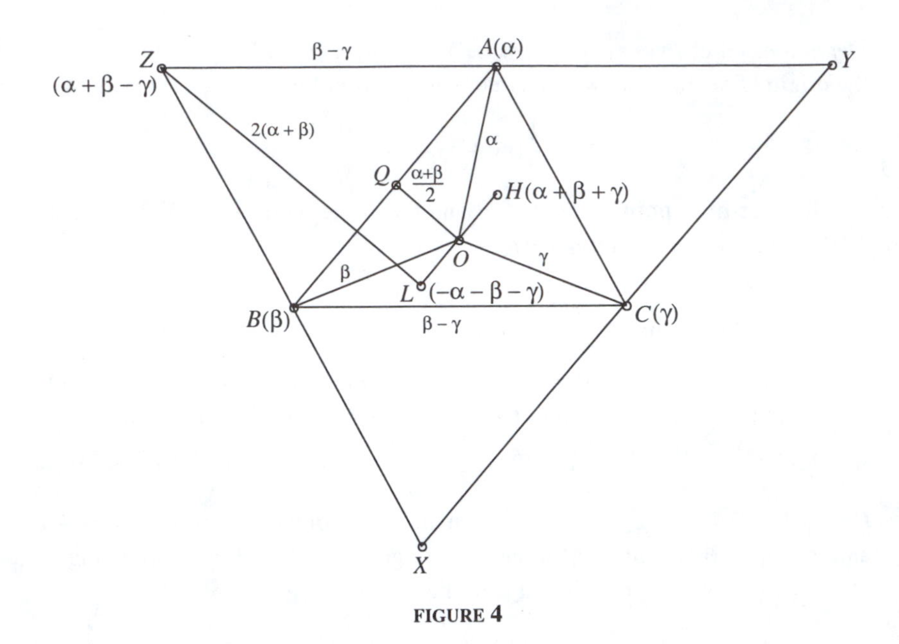

FIGURE 4

(iii) Theorem: *The de Longchamps point of a triangle is the **orthocenter of the anticomplementary triangle**.*

Taking O as the origin of vectors and letting the vectors to the vertices respectively be α, β, and γ, the orthocenter H is given by $(\alpha+\beta+\gamma)$, and hence L would be $-(\alpha+\beta+\gamma)$. Now, the vectors **AZ** and **CB** are equal and parallel, and so $\mathbf{AZ}=\mathbf{CB}=\beta-\gamma$. Thus Z is the point $(\alpha+\beta-\gamma)$, and

$$\mathbf{LZ}=(\alpha+\beta-\gamma)-[-(\alpha+\beta+\gamma]=2(\alpha+\beta)=4\cdot\left[\frac{\alpha+\beta}{2}\right]=4\cdot\mathbf{OQ},$$

where Q is the midpoint of AB.

Thus LZ is parallel to OQ, which is perpendicular to AB, and hence also perpendicular to XY (because XY is parallel to AB), making LZ perpendicular to XY and placing L on the altitude of ΔXYZ from Z. Similarly, L is on the other altitudes and is the orthocenter of ΔXYZ.

Professor Liang-shin Hahn (University of New Mexico), has observed that this property of the de Longchamps point follows immediately from the fact, established above (Figure 2), that the nine-point center N of a triangle is the midpoint of the segment HO joining the orthocenter and the circumcenter. One need only note that the orthocenter H and circumcenter O of a triangle are also respectively the circumcenter and nine-point center of its anticomplementary triangle:

> that is to say, in Figure 4, H and O are respectively the circumcenter (O) and nine-point center (N) of $\triangle XYZ$. Thus, while extending HO its own length in triangle ABC gives L, in triangle XYZ it amounts to extending ON its own length and hence gives its orthocenter H (Figure 5). To see this, observe in Figure 4 that altitude CH of $\triangle ABC$ is perpendicular to AB and therefore also to XY, and since C is the midpoint of XY, CH is in fact the perpendicular bisector of XY; similarly AH and BH are the perpendicular bisectors of YZ and ZX, making H the circumcenter (O) of $\triangle XYZ$. Finally, since the circumcircle of $\triangle ABC$ is also the nine-point circle of $\triangle XYZ$, its center O is the nine- point center (N) of $\triangle XYZ$.

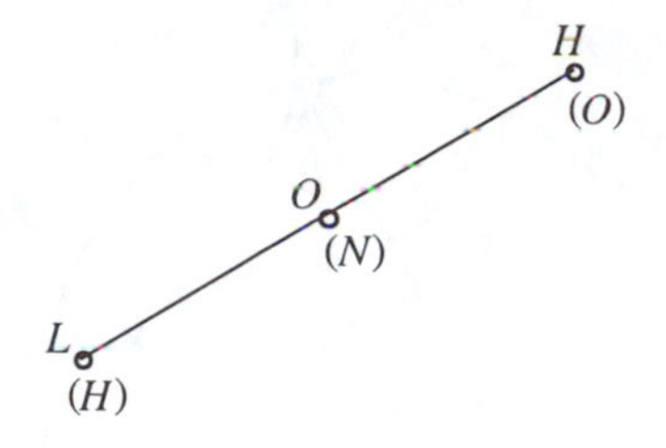

FIGURE 5

4. Cantor's Theorem (M. B. Cantor, 1829–1920)

(a) Let the tangents to the circumcircle of $\triangle ABC$ be drawn at the vertices. Then, if a straight line is drawn from the midpoint of each side of the triangle perpendicular to the tangent at the opposite vertex, the three perpendiculars are concurrent (Figure 6).

Proof: With O as the origin of vectors and the vectors to the vertices respectively α, β, and γ, we have in Figure 7 that the

midpoint A' of $BC = \frac{1}{2}(\beta+\gamma)$, and the nine-point center $N = \frac{1}{2}(\alpha+\beta+\gamma)$.

Hence the vector

$$A'N = \frac{1}{2}(\alpha+\beta+\gamma) - \frac{1}{2}(\beta+\gamma) = \frac{1}{2}\alpha,$$

implying $A'N$ is parallel to α, which is the radius OA, making it perpendicular to the tangent at A. Thus the perpendicular from A' to the tangent at A starts

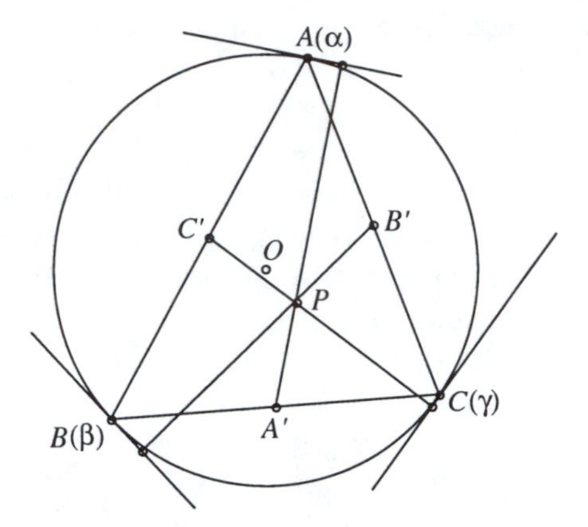

FIGURE 6

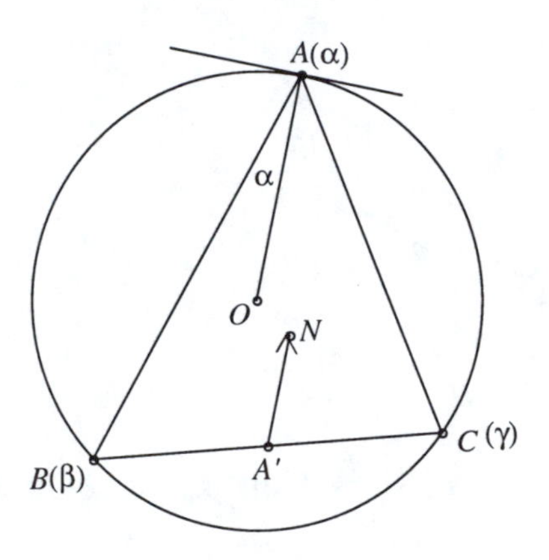

FIGURE 7

out along the segment $A'N$. Similarly the other perpendiculars also go through N, and the proof is complete.

(b) It was very astute of Cantor to identify the midpoint of a side of a triangle as the **centroid** of the vertices on that side, and he made the following general discovery.

Cantor's Theorem

Let $\{z_1, z_2, \ldots, z_n\}$ be a set of n points on a circle. Suppose the tangent is drawn at one of these points and that the perpendicular to this tangent is drawn from the centroid of the other $n-1$ points of the set. If this is done for a tangent at each z_i, the n perpendiculars are always concurrent.

(i) A Preliminary Discussion on Centroids

The centroid of a set of points $S = \{z_1, z_2, \ldots, z_n\}$ is the center of gravity of a system of unit masses, one at each z_i. We shall show that the centroid of $S = \{z_1, z_2, \ldots, z_n\}$ is $\mathbf{G}_n = \frac{1}{n}(z_1+z_2+\cdots+z_n)$, for **all** choices of the origin of vectors.

Proof: (Induction) For $n = 3$, $S = \{z_1, z_2, z_3\}$.

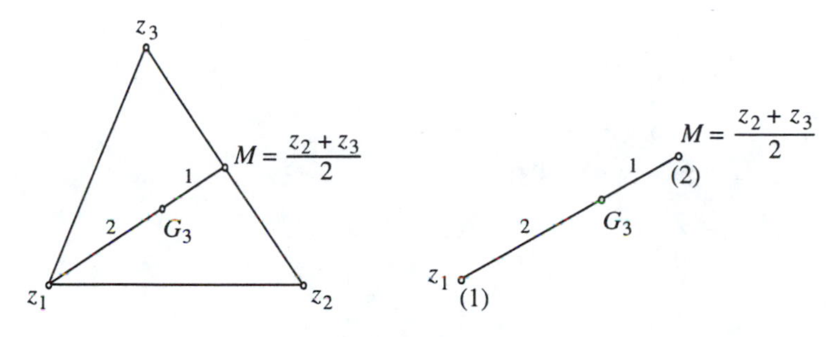

FIGURE 8

The centroid of z_2z_3 is $M = \frac{1}{2}(z_2 + z_3)$, and so a system of unit masses at z_1, z_2, and z_3 is equivalent to a mass of 2 at M and a mass of 1 at z_1. Thus the centroid $\mathbf{G}_3$ divides the segment z_1M in the ratio of 2:1 (Figure 8). That is,

$$\begin{aligned} 2 \cdot \mathbf{MG}_3 &= \mathbf{G}_3 z_1, \\ 2\left[\mathbf{G}_3 - \frac{1}{2}(z_2 + z_3)\right] &= z_1 - \mathbf{G}_3, \\ 2 \cdot \mathbf{G}_3 - (z_2 + z_3) &= z_1 - \mathbf{G}_3, \\ 3 \cdot \mathbf{G}_3 &= z_1 + z_2 + z_3, \end{aligned}$$

giving

$$\mathbf{G}_3 = \frac{1}{3}(z_1 + z_2 + z_3).$$

Suppose the formula holds for a set of $n - 1$ points, and that $S = \{z_1, z_2, \ldots, z_n\}$. Let the centroid of $S - z_n$ be

$$\mathbf{G}_{n-1} = \frac{1}{n-1}(z_1 + z_2 + \cdots + z_{n-1}).$$

Then $\mathbf{G}_n$ divides $\mathbf{G}_{n-1}z_n$ in the ratio $1 : (n - 1)$ (Figure 9).

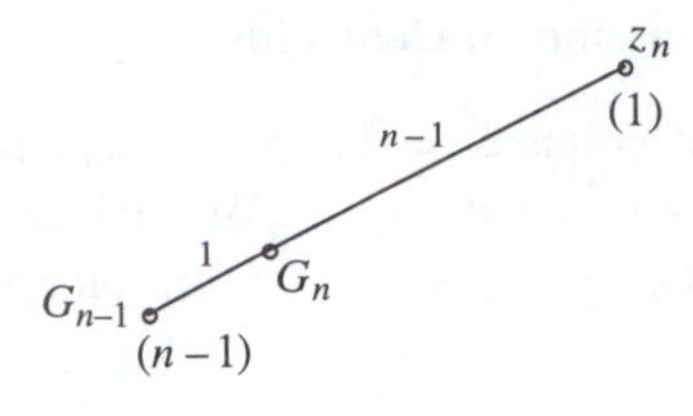

FIGURE 9

Hence

$$(n-1)[\mathbf{G}_n - \mathbf{G}_{n-1}] = z_n - \mathbf{G}_n,$$

giving

$$\begin{aligned} n\mathbf{G}_n &= (n-1)\mathbf{G}_{n-1} + z_n \\ &= (n-1)\frac{1}{n-1}(z_1 + z_2 + \cdots + z_{n-1}) + z_n \\ &= z_1 + z_2 + \cdots + z_{n-1} + z_n, \end{aligned}$$

implying

$$\mathbf{G}_n = \frac{1}{n}(z_1 + z_2 + \cdots + z_{n-1} + z_n), \text{ as desired.}$$

(ii) Proof of Cantor's Theorem

Let the center of the circle be the origin of vectors. Let $z_1 + z_2 + \cdots + z_n = \mathbf{T}$, and consider the point X given by $\frac{\mathbf{T}}{n-1}$. Let $S_i = \{S - z_i\}$ and let $\mathbf{G}_i$ be the centroid of S_i. Then

$$\mathbf{G}_i = \frac{1}{n-1}(\mathbf{T} - z_i),$$

and the vector

$$\mathbf{G}_i\mathbf{X} = \mathbf{X} - \mathbf{G}_i = \frac{z_i}{n-1},$$

implying $\mathbf{G}_i\mathbf{X}$ is parallel to the radius z_i and hence perpendicular to the tangent at z_i. Therefore all the perpendiculars are concurrent at X.

(c) Napoleon's Theorem

If equilateral triangles PAB, QAC, and RBC are drawn outwardly on the sides of $\triangle ABC$, their centers S, T, and U determine another equilateral triangle (Figure 10).

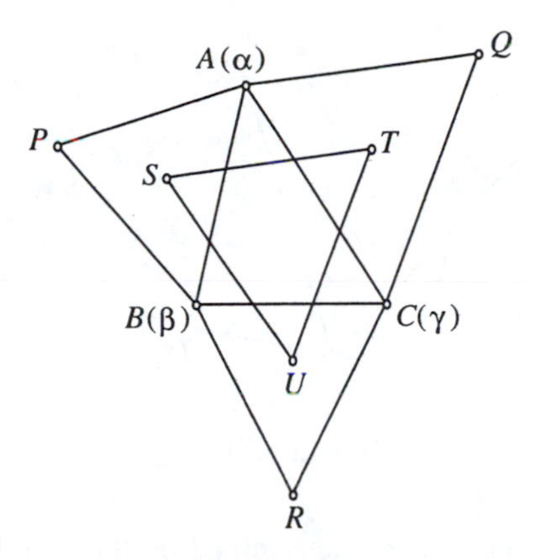

FIGURE 10

(i) To begin, let us devise a way to test the coordinate vectors of three points to determine whether they form an equilateral triangle.

To this end, recall the three cube roots of unity, 1 and $\frac{1\pm\sqrt{-3}}{2}$. Letting $\frac{1+\sqrt{-3}}{2} = \omega$, it is a simple calculation to show that $\frac{-1-\sqrt{-3}}{2} = \omega^2$, and since the sum of the roots of $x^3 - 1 = 0$, we have

$$1 + \omega + \omega^2 = 0.$$

Now, all roots of unity lie on the unit circle $|z| = 1$, and for the cube root ω, it is evident that

$$\arg \omega = \frac{2\pi}{3} \quad \text{(i.e., } 120°).$$

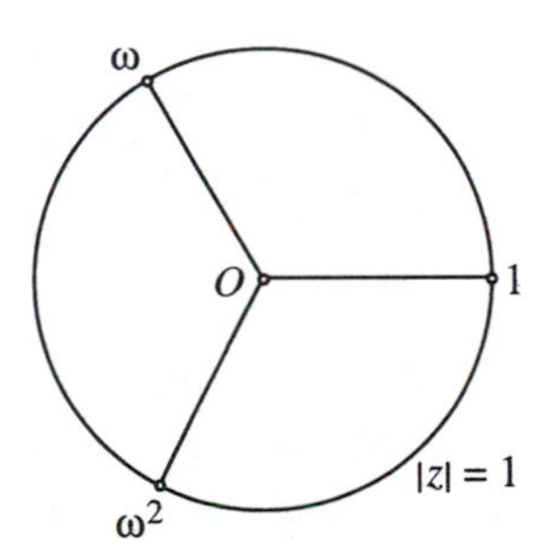

FIGURE 11

Accordingly, the effect of multiplying a vector by ω is merely to rotate it through the angle $\frac{2\pi}{3}$ while leaving its magnitude unchanged.

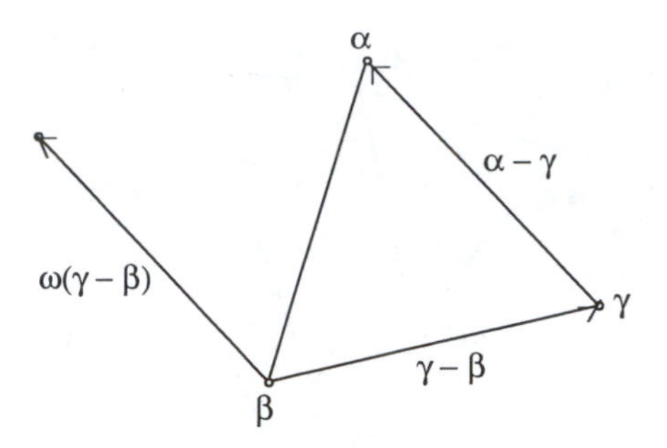

FIGURE 12

Consider, then, the triangle determined by the coordinate vectors α, β, and γ, where the cyclic order $\alpha \to \beta \to \gamma$ is **counterclockwise**. Then, as shown in Figure 12, the vector from the endpoint of β to the endpoint of γ is $\gamma - \beta$ and the vector from the endpoint of γ to the endpoint of α is $\alpha - \gamma$. Now, the triangle is equilateral if and only if rotating this vector $\gamma - \beta$ about the point β through the angle $\frac{2\pi}{3}$ makes it equal and parallel to $\alpha - \gamma$; that is, iff

$$\omega(\gamma - \beta) = \alpha - \gamma,$$

or

$$(1 + \omega)\gamma - \omega\beta = \alpha.$$

But $1 + \omega = -\omega^2$, and the condition is therefore

$$-\omega^2\gamma - \omega\beta = \alpha,$$

i.e.,

$$\alpha + \omega\beta + \omega^2\gamma = 0.$$

That is to say, multiply the coordinate vectors of the vertices in counterclockwise order by the cube roots of unity in the order 1, ω, ω^2; the triangle is equilateral if and only if the sum of these products is zero. (If the cyclic order $\alpha \to \beta \to \gamma$ is clockwise, the ω and ω^2 need to be interchanged.)

(ii) Now let us prove Napoleon's theorem. Since $\triangle PAB$ is equilateral, it follows that

$$P + \omega\beta + \omega^2\alpha = 0; \tag{1}$$

Similarly, triangles QAC and RBC yield:

$$R + \omega\gamma + \omega^2\beta = 0, \tag{2}$$

and

$$Q + \omega\alpha + \omega^2\gamma = 0. \tag{3}$$

Now, S, T, U are the centers of their equilateral triangles, that is, their **centroids**, and therefore we have that

$$S = \frac{1}{3}(P + \alpha + \beta), \quad T = \frac{1}{3}(Q + \gamma + \alpha), \quad U = \frac{1}{3}(R + \beta + \gamma).$$

To test $\triangle STU$, we need to determine whether $S + \omega U + \omega^2 T$ is equal to 0. We have

$$\begin{aligned} S &+ \omega U + \omega^2 T \\ &= \frac{1}{3}[P + \alpha + \beta + \omega R + \omega\beta + \omega\gamma + \omega^2 Q + \omega^2\gamma + \omega^2\alpha] \\ &= \frac{1}{3}[1 \cdot \text{(the left side of equation 1)} + \omega \cdot \text{(the left side of equation 2)} \\ &\qquad + \omega^2 \cdot \text{(the left side of equation 3)}] \\ &= 0, \end{aligned}$$

implying $\triangle STU$ is indeed equilateral.

Reference

1. Clark Kimberling, "Central Points and Central Lines in the Plane of a Triangle," *Mathematics Magazine*, June 1994, 163–187.

SECTION 23

A Problem from the Philippines

(Unused from the 1991 International Olympiad)

> From a point P in the interior of $\triangle ABC$, perpendiculars PQ and PR are dropped to the sides BC and AC (Figure 1). Also, from vertex C, perpendiculars CS and CT are drawn to the extensions of AP and BP. Prove the engaging fact that, wherever the point P might be taken inside $\triangle ABC$, the point of intersection U of SQ and TR always lies on AB.

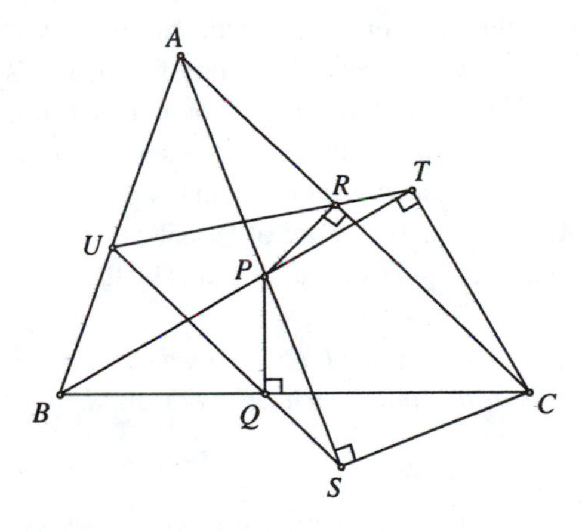

FIGURE 1

The right angles imply that the circle on diameter PC goes through the six points P, Q, S, C, T, R (Figure 2). That is to say, $PQSCTR$ is a hexagon inscribed in a circle. By Pascal's famous theorem of the "mystic" hexagon, the three points of intersection determined by the three pairs of opposite sides of such a hexagon are collinear, no matter how the vertices might be ordered to specify the sides of the hexagon. For the hexagon $PSQCRT$, then, the pairs of opposite sides are

$$(PS, CR), (SQ, RT), (QC, TP),$$

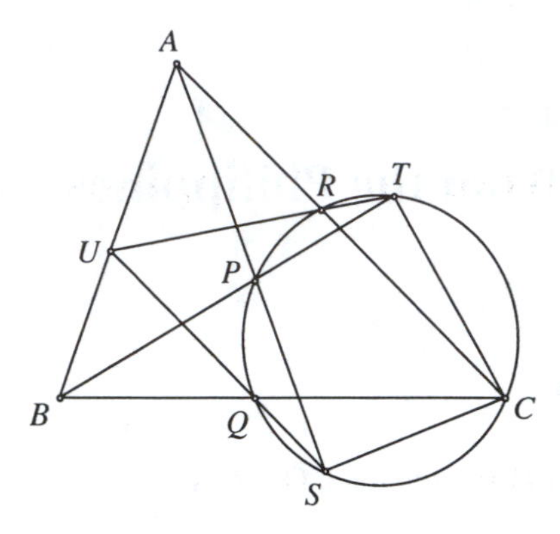

FIGURE 2

making the corresponding collinear points A, U, and B, and the problem is solved already.

If this problem had been used on the olympiad, I suppose the examiners would have accepted this solution, but it seems more likely that the proposers expected an argument that is essentially a proof of Pascal's theorem. For a candidate who is not familiar with Pascal's theorem and its proof, this would be quite a difficult problem. However, the contestants in an international olympiad are so inventive that it is not at all beyond them to approach the problem along the lines of the following standard proof of Pascal's theorem.

One way to show three points are collinear is by the theorem of Menelaus:

three points, one on each side of a triangle, are collinear if and only if the product K of the ratios into which the sides are divided by the points is -1.

To avail ourselves of this theorem we need to identify a triangle XYZ which has one of A, U, and B on each of its sides. It is far from obvious that a suitable choice is obtained by taking PS as the side through A, QC as the side through B, and RT as the side through U (Figure 3). In this case, the other principal lines through A, U, and B, namely CRA, SQU, and TPB, are transversals of $\triangle XYZ$ and they can be used to invoke the converse part of Menelaus' theorem. Just before doing that, however, let us observe that our goal is to show that the product of ratios

$$K = \frac{XU}{UZ} \cdot \frac{ZB}{BY} \cdot \frac{YA}{AX} = -1.$$

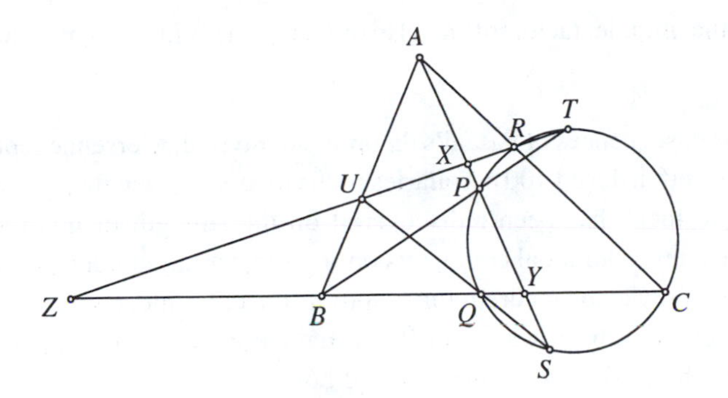

FIGURE 3

Now, since CRA is a transversal of $\triangle XYZ$, we have

$$\frac{XR}{RZ} \cdot \frac{ZC}{CY} \cdot \frac{YA}{AX} = -1, \quad \text{giving the ratio} \quad \frac{YA}{AX} = -\frac{RZ \cdot CY}{XR \cdot ZC};$$

similarly, the transversals SQU and TPB yield

$$\frac{XU}{UZ} \cdot \frac{ZQ}{QY} \cdot \frac{YS}{SX} = -1, \quad \text{implying} \quad \frac{XU}{UZ} = -\frac{QY \cdot SX}{ZQ \cdot YS},$$

and

$$\frac{XT}{TZ} \cdot \frac{ZB}{BY} \cdot \frac{YP}{PX} = -1, \quad \text{giving} \quad \frac{ZB}{BY} = -\frac{TZ \cdot PX}{XT \cdot YP}.$$

Hence

$$K = -\frac{QY \cdot SX}{ZQ \cdot YS} \cdot \frac{TZ \cdot PX}{XT \cdot YP} \cdot \frac{RZ \cdot CY}{XR \cdot ZC},$$

and, grouping together the factors containing respectively X, Y, and Z, we get

$$K = -\left(\frac{SX \cdot PX}{XT \cdot XR}\right) \cdot \left(\frac{QY \cdot CY}{YS \cdot YP}\right) \cdot \left(\frac{TZ \cdot RZ}{ZQ \cdot ZC}\right).$$

But, since XPS and XRT are secants to the circle from the same point X, we have

$$XT \cdot XR = XS \cdot XP = SX \cdot PX,$$

revealing that the first factor in this expression for K is simply 1. Similarly, the secants ZQC and ZRT show that the third factor in K is also 1. Finally, since the chords PS and QC of the circle intersect at Y, we have

$$QY \cdot YC = PY \cdot YS, \quad \text{i.e.,} \quad QY \cdot CY = YP \cdot YS,$$

making the middle factor of K also equal to 1, and we have the desired $K = -1$.

The consequences of Pascal's theorem are myriad; Mersenne reported that Pascal himself deduced 400 corollaries to his theorem. Even though all of these are lost, so much has been rediscovered on the strength of this remarkable result that a few additional comments on the subject might not be out of place.

Pascal proved the theorem for six points on any conic, not just circles, and his essay on the subject was so brilliant that Descartes refused to believe that it could be the work of a sixteen year old boy!

(a) The line through the three collinear points of Pascal's theorem is called **the Pascal line** of the hexagon from which it is determined. Since there are 60 ways of ordering six points around a conic, the same six points give rise to 60 different inscribed hexagons and a corresponding set of 60 Pascal lines. Since each Pascal line is determined by three points of intersection, the 60 lines hold a total of 180 "defining" points altogether. However, since a pair of opposite sides of a hexagon retains its status as a pair of opposite sides through four ways of ordering the sides, each defining point contributes to the determination of four Pascal lines:

> for example, for the six points A, B, C, D, E, F, the sides AB and DE are opposite for the four hexagons $ABCDEF$, $ABCEDF$, $ABFDEC$, and $ABFEDC$.

Thus each defining point counts four toward this total of 180 line-point incidences, implying that there are only $\frac{180}{4} = 45$ different defining points. That is to say,

> **the 60 Pascal lines intersect in 45 points such that (i) each Pascal line passes through 3 points, and (ii) through each point there are 4 Pascal lines (Figure 4).**

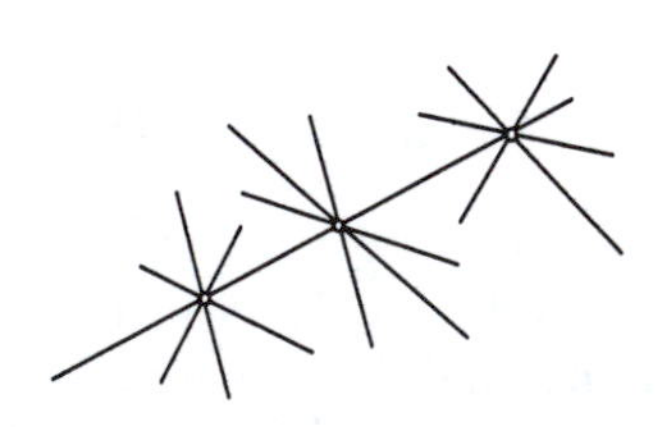

FIGURE 4

(b) In general, the 60 Pascal lines provide $\frac{60 \cdot 59}{2} = 1770$ points of intersection, 45 of which are the aforementioned "defining" points. Another 20 of these points, which lie one on each Pascal line, are called Steiner points (after the great Swiss geometer Jacob Steiner, 1796–1863). The 60 Pascal lines, then, hold a total of 60 line-point incidences, but because there are only 20 different Steiner points, it must be that each Steiner point is counted three times in this total. Thus there exists

> **a set of 20 points, called Steiner points, such that (i) each Pascal line passes through a single Steiner point and (ii) through each Steiner point there are three Pascal lines (Figure 5).**

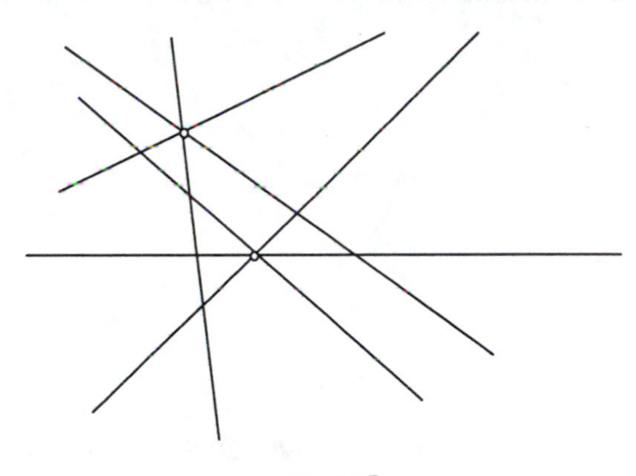

FIGURE 5

(c) Now erase everything in the figure except the 20 Steiner points. Each of the $\frac{20 \cdot 19}{2} = 190$ pairs of Steiner points determines a straight line, a particular 15 of which are called Plücker lines (after the distinguished German geometer-physicist Julius Plücker, 1801–1868). Each Plücker line goes through four Steiner points, for a total of 60 incidences. Thus each of the 20 Steiner points must count three towards this total, with the result that

> **the 20 Steiner points determine 15 Plücker lines such that (i) each Plücker line passes through four Steiner points and (ii) through each Steiner point there are three Plücker lines (Figure 6).**

(d) Reinstating the Pascal lines, there is, among their points of intersection, another set of **60 Kirkman points** (named after the British mathematician Thomas Kirkman, 1806–1895), such that

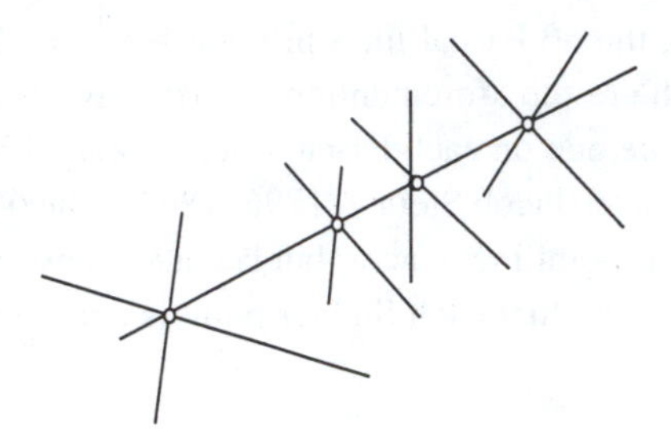

FIGURE 6

(i) each Pascal line passes through three Kirkman points and (ii) through each Kirkman point there are three Pascal lines (Figure 7).

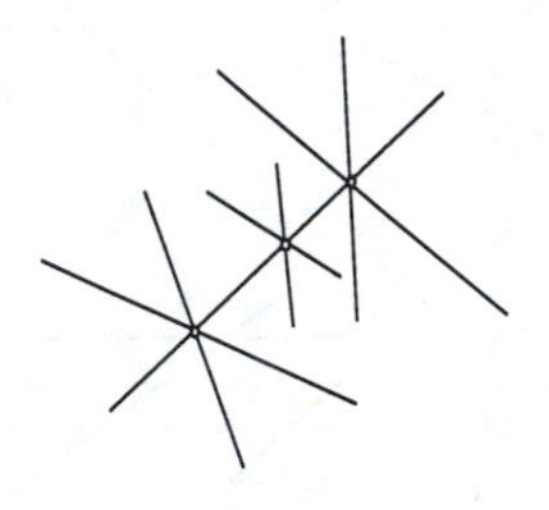

FIGURE 7

(e) Now, the 60 Kirkman points and the 20 Steiner points comprise a set of 80 points which generate **20 Cayley lines** (named after the great English mathematician Arthur Cayley, 1821–1895) such that

(i) each Cayley line contains three Kirkman points and one Steiner point, and (ii) each Kirkman point and each Steiner point lies on a unique Cayley line (Figure 8).

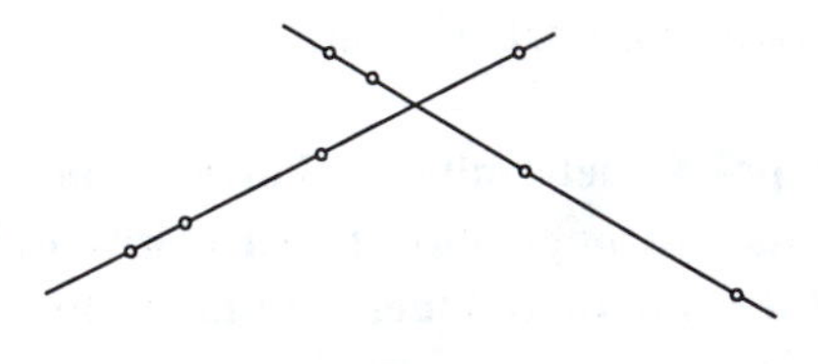

FIGURE 8

(f) Finally, in honor of the Dublin mathematician George Salmon (1819–1904),

the 20 Cayley lines intersect in 15 Salmon points such that (i) each Cayley line passes through three Salmon points and (ii) through each Salmon point there are four Cayley lines (Figure 9).

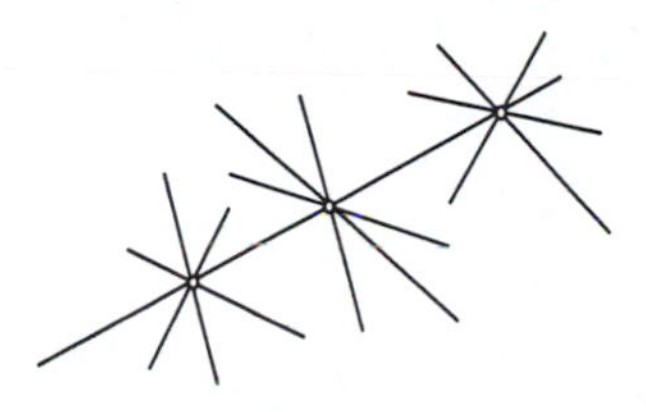

FIGURE 9

These observations hardly scratch the surface of this fertile configuration. These and many more discoveries belong to the nineteenth century and it is almost inconceivable that additional results have not been found in the twentieth century. For further information on the state of the subject around the turn of the century, see Salmon's *Conic Sections*, pages 379–383, where advances by Veronese, Cremona, and Cathcart are reported. Another reference of the same period is Lachlan's *Modern Pure Geometry*.

SECTION 24

Four Solutions by George Evagelopoulos

The law student George Evagelopoulos will not be a stranger to readers of *From Erdös to Kiev*, which contains ten examples of his ingenuity. George has now completed his studies and at present (1995) is practising criminal law in Athens, Greece.

George is keeping his hand in at mathematics and recently he sent me the following solutions to four problems from the *American Mathematical Monthly*. He is also the Editor-in-Chief of the Greek edition of the outstanding journal *Quantum*.

1. Problem 10259, November 1992, page 873; proposed by Jonathan L. King, University of Florida, Gainesville, Florida.

   The sequence $R = \{r_0, r_1, r_2, \ldots\} = \{3, 7, 47, 2207, \ldots\}$ begins with $r_0 = 3$ and thereafter has $r_n = r_{n-1}^2 - 2$. From R, the kth term of a second sequence S is determined from the product of the first k terms of R by taking the square root once for each term in the product:

$$S = \left\{ \sqrt{3}, \sqrt{\sqrt{3 \cdot 7}}, \sqrt{\sqrt{\sqrt{3 \cdot 7 \cdot 47}}}, \ldots \right\}$$

   To what limit does S converge?

Clearly the desired limit is given by

$$\lim_{k \to \infty} (r_0 r_1 \ldots r_{k-1})^{1/2^k}.$$

Many solutions begin with the asking of an appropriate question, and it appears that George wondered about the form of r_{n-1} in order that its square might generate a term $+2$ to cancel the -2 in the formula for r_n. Having asked the question, it is not a big step to the realization that

$$\left(a + \frac{1}{a}\right)^2 = a^2 + 2 + \frac{1}{a^2},$$

with the implication that

$$\text{if} \quad r_{n-1} = a + \frac{1}{a}, \quad \text{then} \quad r_n = a^2 + \frac{1}{a^2},$$

which in turn yields

$$r_{n+1} = a^4 + \frac{1}{a^4},$$

and so on. Thus his solution begins by setting

$$r_0 = 3 = a + \frac{1}{a},$$

giving

$$r_1 = a^2 + \frac{1}{a^2},$$

and in general

$$r_k = a^{2^k} + \frac{1}{a^{2^k}} = a^{2^k} + a^{-2^k}.$$

Solving

$$3 = a + \frac{1}{a},$$

we get

$$a^2 - 3a + 1 = 0, \quad \text{and} \quad a = \frac{3 \pm \sqrt{5}}{2}.$$

Either value of a generates R, and for definiteness George takes the value that is greater than 1,

$$a = \frac{3 + \sqrt{5}}{2}.$$

With a known, and $r_k = a^{2^k} + a^{-2^k}$, the next step would appear to be to write

$$(r_0 r_1 \ldots r_{k-1})^{1/2^k} \quad \text{as} \quad \left[(a + a^{-1})(a^2 + a^{-2}) \cdots (a^{2^{k-1}} + a^{-2^{k-1}})\right]^{1/2^k}$$

However, this soon leads to an impasse and George cleverly observes instead that $r_k = a^{2^k} + a^{-2^k}$ can be used to establish the inequalities

$$a^{2^k} < r_k < 2a^{2^k} :$$

since $a > 1$, we have

$$a^{2^k} < a^{2^k} + a^{-2^k} = r_k < a^{2^k} + 1 < 2a^{2^k}.$$

It's all downhill from here, for now we have

$$a^{2^0} \cdot a^{2^1} \cdots a^{2^{k-1}} < r_0 r_1 \cdots r_{k-1} < 2^k \cdot a^{2^0} \cdot a^{2^1} \cdots a^{2^{k-1}},$$

that is,

$$a^{1+2+2^2+\cdots+2^{k-1}} < r_0 r_1 \cdots r_{k-1} < 2^k \cdot a^{1+2+2^2+\cdots+2^{k-1}},$$
$$a^{2^k-1} < r_0 r_1 \cdots r_{k-1} < 2^k \cdot a^{2^k-1},$$

and

$$a^{1-(1/2^k)} < (r_0 r_1 \cdots r_{k-1})^{1/2^k} < 2^{k/2^k} \cdot a^{1-(1/2^k)}.$$

Since $\frac{k}{2^k} \to 0$ as $k \to \infty$, the desired limit is simply a itself, $\frac{3+\sqrt{5}}{2}$, which we observe is the square of the ubiquitous golden ratio $\alpha = \frac{1+\sqrt{5}}{2}$.

2. Now for a lovely problem proposed by Donald Knuth, Stanford University, Stanford, California: Problem 10280, January 1993, page 76.

   A binary operation $*$ on the set $S = \{1, 2, \ldots, n\}$ is a mapping of the n^2 pairs of elements of S, $\{(1, 1), (1, 2), \ldots, (n, n)\}$, to S itself. We write $a * b = c$ to indicate that the pair (a, b) is mapped to c.

   Each of the n possible values of $1 * 1$ can be combined with each of the n possibilities for $1 * 2$, and so on to the n choices for $n * n$, for a grand total of

$$n \cdot n \cdot n \cdot \cdots \cdot n = n^{n^2}$$

   binary operations on S.

   Let x be a member of S and suppose that the operation $*$ yields $x * x = a$, $a * x = b$, and $b * b = c$. These mappings mark the steps involved in evaluating the expression $[(x * x) * x] * [(x * x) * x]$:

$$\begin{aligned}[(x * x) * x] * [(x * x) * x] &= (a * x) * (a * x) \\ &= b * b \\ &= c.\end{aligned}$$

   Now, there are some binary operations $*$ which return the number x as this result c. For some operations, $c = x$ might hold for certain values of x and fail for others. Obviously it is rarer for an operation to return $c = x$ for **every** $x = 1, 2, \ldots, n$. But it is just this subset A of the binary operations on S that is the subject of this problem.

(a) If $*$ is a binary operation on S that is chosen at random from the n^{n^2} possible such operations, prove that the probability it belongs to A is

$$P_n = \sum_{k=0}^{n} \frac{p(n,k)}{n^{2n-k}},$$

where $p(n,k)$ is the number of permutations of $(1, 2, \ldots, n)$ which have k fixed elements.

(b) Secondly, prove that, as $n \to \infty$, P_n is asymptotic to $\frac{e^{n-1}n!}{2n^{2n}}$, that is,

$$\text{the ratio } \frac{P_n}{\frac{e^{n-1}n!}{2n^{2n}}} \to 1 \text{ as } n \to \infty.$$

George's solution of part (b) is quite sophisticated, and so let us content ourselves with his brilliant solution of part (a).

The problem is straightforward enough—determine the number of operations in A and divide by n^{n^2}.

If $[(x * x) * x] * [(x * x) * x] = x$ for all x in S, let us say that the operation $*$ has property P. Thus each $*$ in A has property P and George observes that these operations can be classified according to the way they map the expressions $(x * x) * x$ as x runs through the members of S. Let $*$ belong to A and let y and z be two different members of S. If $*$ gives the same result for

$$(y * y) * y \quad \text{and} \quad (z * z) * z,$$

then by property P we would have the contradiction

$$y = [(y * y) * y] * [(y * y) * y] = [(z * z) * z] * [(z * z) * z] = z.$$

Thus $(y * y) * y$ and $(z * z) * z$ must be different when y and z are different. Consequently, as x runs through $1, 2, \ldots, n$, the values of $(x * x) * x$, being all different, are again the numbers $1, 2, \ldots, n$ in some order. That is to say, $(x * x) * x$ generates a permutation σ of S. Therefore, let the operations be classified according to their permutations, collecting those that generate σ in a subset A_σ.

Now, let σ be any permutation of $(1, 2, \ldots, n)$. We have just stated that the qualifications required of an operation $*$ for membership in A_σ are

(i) $[(x * x) * x] * [(x * x) * x] = x$ for all x in S (property P), and

(ii) $(x * x) * x = \sigma(x)$ for all x in S.

George now makes the inspired observation that these requirements are equivalent to the following necessary and sufficient condition: $*$ belongs to A_σ if and only if

(iii) $\sigma(x) * \sigma(x) = x$ for all x in S, and
(iv) $x * \sigma(x) = \sigma^2(x)$ for all x in S, where $\sigma^2(x)$ stands for $\sigma(\sigma(x))$.

Proof: Necessity: Suppose $*$ belongs to A_σ, that is, $*$ satisfies (i) and (ii).

Then, by (i),

$$[(x * x) * x] * [(x * x) * x] = x,$$

which, by (ii), is just

$$\sigma(x) \cdot \sigma(x) = x,$$

which is (iii).

Also, applying (ii) to $\sigma(x)$ itself, we get

$$(\sigma(x) * \sigma(x)) * \sigma(x) = \sigma(\sigma(x)),$$

which, in view of (iii) (which was just established) is (iv):

$$x * \sigma(x) = \sigma^2(x).$$

Sufficiency: Suppose $*$ satisfies (iii) and (iv).

First let us establish (ii). Applying (iii) to the inverse $\sigma^{-1}(x)$, we obtain

$$\sigma(\sigma^{-1}(x)) * \sigma(\sigma^{-1}(x)) = \sigma^{-1}(x),$$

which is just

$$x * x = \sigma^{-1}(x).$$

Next, applying (iv) to $\sigma^{-1}(x)$, we have

$$\sigma^{-1}(x) * \sigma(\sigma^{-1}(x)) = \sigma^2(\sigma^{-1}(x)),$$

which reduces to

$$\sigma^{-1}(x) * x = \sigma(x).$$

Thus (ii) follows directly:

$$(x * x) * x = \sigma^{-1}(x) * x = \sigma(x).$$

Finally, (i) follows from (ii) and (iii): since

$$(x * x) * x = \sigma(x) \quad \text{and} \quad \sigma(x) * \sigma(x) = x,$$

we immediately have

$$[(x * x) * x] * [(x * x) * x] = x.$$

This argument also brings out the fact that each $*$ in A_σ must map the pair (x, x) to $\sigma^{-1}(x)$ for all x in S: as we observed above, applying (iii) to $\sigma^{-1}(x)$ gives

$$x * x = \sigma^{-1}(x).$$

That is to say, every operation in A_σ assigns the same n images to the n pairs $(1, 1), (2, 2), \ldots, (n, n)$.

Now, if σ has k fixed points and x is one of the $n - k$ values of S which is **not** a fixed point, then an operation $*$ in A_σ has no option but to map the pair $(x, \sigma(x))$ to $\sigma^2(x)$ in order to satisfy condition (iv): $x * \sigma(x) = \sigma^2(x)$. Thus each $*$ in A_σ maps $(x, \sigma(x))$ to the same image, $\sigma^2(x)$. Now, if x is not a fixed point of σ, then $\sigma(x)$ cannot be x itself, making the pair $(x, \sigma(x))$ **different** from any of the n pairs (x, x). In summary, then, if σ has k fixed points, each $*$ in A_σ must map the n pairs (x, x) and the $n - k$ pairs $(x, \sigma(x))$ to the same set of $2n - k$ images.

Conversely, assigning these $2n - k$ images as just prescribed is all an operation $*$ has to do in order to satisfy (iii) and (iv) and thus qualify for membership in A_σ:

(1) If $x * x = \sigma^{-1}(x)$ for the n pairs (x, x), then it holds for the pair $(\sigma(x), \sigma(x))$, satisfying (iii):

$$\sigma(x) * \sigma(x) = \sigma^{-1}(\sigma(x)) = x.$$

(2) Now, if f is a fixed value of σ, then

$$\begin{aligned} f * \sigma(f) = f * f = \sigma^{-1}(f) &\quad \text{by the rule governing pairs } (x, x) \\ = f = \sigma^2(f), &\quad \text{since } f \text{ is a fixed point,} \end{aligned}$$

i.e., (iv) holds for $x = f$:

$$f * \sigma(f) = \sigma^2(f);$$

hence, prescribing $x * \sigma(x) = \sigma^2(x)$ for the $n - k$ values x which are not fixed points of σ ensures condition (iv) for **all** x in S.

Thus every $*$ which contains these $2n - k$ obligatory images belongs to A_σ. Whichever of the n values $(1, 2, \ldots, n)$ the operation might assign to each of the remaining $n^2 - 2n + k$ pairs (x, y) is irrelevant. Therefore there are n^{n^2-2n+k} different ways of assigning the remaining images and hence of constructing an operation $*$ so that it qualifies for membership in A_σ; thus

$$|A_\sigma| = n^{n^2-2n+k} \quad \text{for a permutation } \sigma \text{ with } k \text{ fixed points.}$$

Consequently, the $p(n, k)$ permutations with k fixed points account for a total of

$$p(n, k) \cdot n^{n^2-2n+k}$$

operations in A.

Finally, then,

$$|A| = \sum_{k=0}^{n} p(n, k) \cdot n^{n^2-2n+k},$$

and the probability of a random operation belonging to A is

$$\begin{aligned} P_n &= \frac{1}{n^{n^2}} \sum_{k=0}^{n} p(n, k) \cdot n^{n^2-2n+k} \\ &= \sum_{k=0}^{n} \frac{p(n, k)}{n^{2n-k}}, \end{aligned}$$

as desired.

3. Problem 10292, part (a), March 1993, page 291; proposed by Jean Anglesio, Garches, France.

Determine the sum

$$S = \sum_{n=1}^{\infty} \arctan \frac{2}{n^2}.$$

Recall that $\arctan x - \arctan y = \arctan \frac{x-y}{1+xy}$,

$$\left(\text{the tangent of each side is } \frac{x - y}{1 + xy}\right).$$

In view of this, let us try to express $\arctan \frac{2}{n^2}$ as a difference, in the hope of obtaining a telescoping series. To this end, we need to find x and y such that

$$\frac{2}{n^2} = \frac{x - y}{1 + xy}.$$

Although not generally a successful approach, equating numerators and also equating denominators is most gratifying in this case:

$$2 = x - y \quad \text{and} \quad n^2 = 1 + xy$$

leads to

$$n^2 - 1 = (n+1)(n-1) = xy$$

and the solution

$$x = n+1,\ y = n-1,\ \text{since } n+1-(n-1) = 2.$$

Therefore

$$\arctan\frac{2}{n^2} = \arctan(n+1) - \arctan(n-1),$$

and

$$\begin{aligned} S = \lim_{n\to\infty} \Big[&(\arctan 2 - \arctan 0) + (\arctan 3 - \arctan 1) \\ &+ (\arctan 4 - \arctan 2) + (\arctan 5 - \arctan 3) \\ &\vdots \\ &+ [\arctan n - \arctan(n-2)] + [\arctan(n+1) - \arctan(n-1)]\Big] \\ = \lim_{n\to\infty} &[\arctan n - \arctan 0 + \arctan(n+1) - \arctan 1] \\ = \frac{\pi}{2} &- 0 + \frac{\pi}{2} - \frac{\pi}{4} \\ = \frac{3\pi}{4}&. \end{aligned}$$

4. Problem 10262, November 1992, page 873; proposed by Dean Clark, University of Rhode Island, Kingston, Rhode Island.

   If f_n denotes the nth term of the Fibonacci sequence $\{1, 1, 2, 3, \ldots\}$, determine the sum of the series

$$\begin{aligned} S = \sum_{n=1}^{\infty} \frac{(-1)^n}{f_n f_{n+2}} &= -\frac{1}{f_1 f_3} + \frac{1}{f_2 f_4} - \frac{1}{f_3 f_5} + - \cdots \\ &= -\frac{1}{1\cdot 2} + \frac{1}{1\cdot 3} - \frac{1}{2\cdot 5} + - \cdots. \end{aligned}$$

Since $f_{n+2} = f_{n+1} + f_n$, then $f_{n+1} = f_{n+2} - f_n$ and therefore, multiplying by $\frac{f_{n+2}-f_n}{f_{n+1}}$, which is equal to 1, we have

$$S = \sum_{n=1}^{\infty}(-1)^n \frac{f_{n+2} - f_n}{f_n f_{n+1} f_{n+2}}$$
$$= \sum_{n=1}^{\infty}(-1)^n \left[\frac{1}{f_n f_{n+1}} - \frac{1}{f_{n+1} f_{n+2}}\right]$$
$$= \sum_{n=1}^{\infty} \frac{(-1)^n}{f_n f_{n+1}} + \sum_{n=1}^{\infty} \frac{(-1)^{n+1}}{f_{n+1} f_{n+2}}$$
$$= \sum_{n=1}^{\infty} \frac{(-1)^n}{f_n f_{n+1}} + \sum_{n=2}^{\infty} \frac{(-1)^n}{f_n f_{n+1}}$$

(where the second sum begins at $n = 2$)

$$= \sum_{n=1}^{\infty} \frac{(-1)^n}{f_n f_{n+1}} + \left(\sum_{n=1}^{\infty} \frac{(-1)^n}{f_n f_{n+1}} - \frac{-1}{f_1 f_2}\right)$$

(beginning the second sum at $n = 1$ and subtracting off the term for $n = 1$)

$$= 2\sum_{n=1}^{\infty} \frac{(-1)^n}{f_n f_{n+1}} + 1.$$

Now, it is well known that $f_{n+1}^2 - f_n f_{n+2} = (-1)^n$. Hence

$$S = 1 + 2\sum_{n=1}^{\infty} \frac{f_{n+1}^2 - f_n f_{n+2}}{f_n f_{n+1}}$$
$$= 1 + 2\sum_{n=1}^{\infty} \left(\frac{f_{n+1}}{f_n} - \frac{f_{n+2}}{f_{n+1}}\right),$$

containing the telescoping series whose partial sums are

$$s_n = \left(\frac{f_2}{f_1} - \frac{f_3}{f_2}\right) + \left(\frac{f_3}{f_2} - \frac{f_4}{f_3}\right) + \left(\frac{f_4}{f_3} - \frac{f_5}{f_4}\right) + \cdots + \left(\frac{f_{n+1}}{f_n} - \frac{f_{n+2}}{f_{n+1}}\right)$$
$$= 1 - \frac{f_{n+2}}{f_{n+1}},$$

and whose sum is

$$\lim_{n\to\infty} s_n = 1 - \lim_{n\to\infty} \frac{f_{n+2}}{f_{n+1}} = 1 - \alpha,$$

where α is the golden ratio $\frac{1+\sqrt{5}}{2}$ (this limit is also a result of long standing). Hence

$$S = 1 + 2(1 - \alpha) = 1 + 2 - (1 + \sqrt{5}) = 2 - \sqrt{5}.$$

SECTION 25
A Problem from the 1992 Canadian Olympiad

U and V are points on the sides AB and CD of unit square $ABCD$ (Figure 1). AV and UD intersect at P, and UC and BV intersect at Q. Where should U and V be placed in order to make the area of quadrilateral $PUQV$ a maximum?

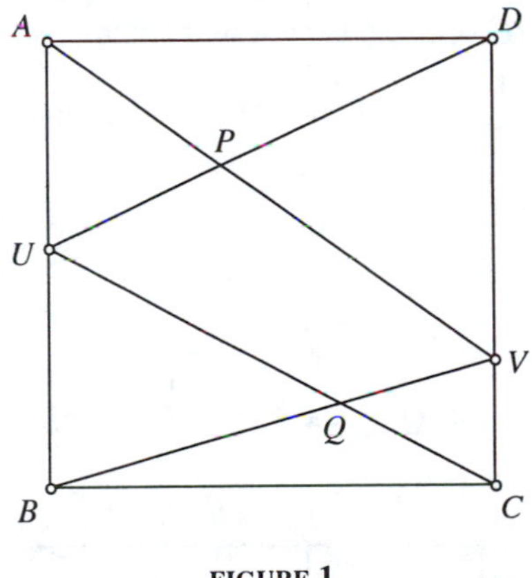

FIGURE 1

The difficulty, of course, is due to the unsymmetrical shape of $PUQV$ and its oblique position in the square. However, things become manageable with the observation that the parts x and y into which the quadrilateral is divided by the diagonal UV have the same areas as triangles APD and BQC: in Figure 2,

$$\Delta AUD = \frac{1}{2} \cdot AU \cdot 1 = \Delta AUV,$$

and subtracting the common region ΔAUP from these triangles, we get

$$\Delta APD = \Delta UPV = x;$$

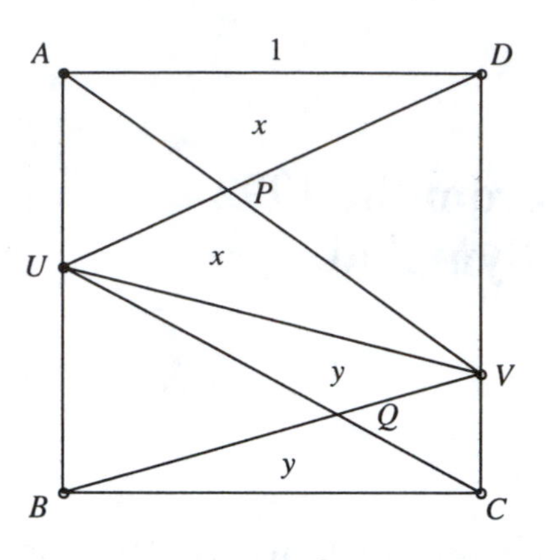

FIGURE 2

similarly,

$$\triangle BQC = \triangle UQV = y.$$

Since the base BC of $\triangle BQC$ is of unit length, its area is simply

$$y = \frac{1}{2} \cdot 1 \cdot (\text{altitude } QE) = \frac{a}{2} \quad \text{(Figure 3)}.$$

Similarly,

$$x = \frac{1}{2} \cdot 1 \cdot b = \frac{b}{2},$$

and

$$PUQV = x + y = \frac{1}{2}(a + b).$$

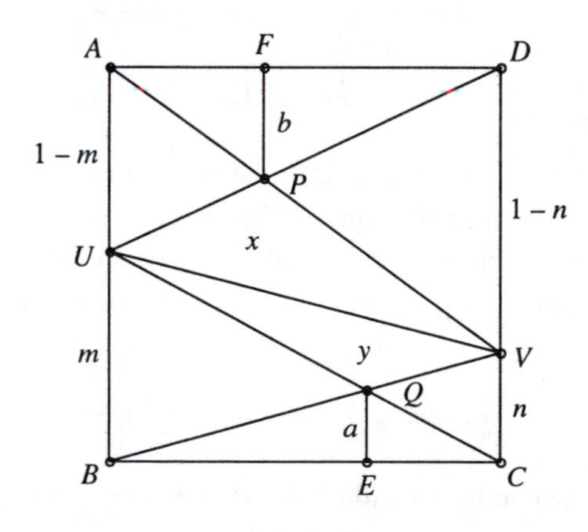

FIGURE 3

Now, because QE is parallel to the sides of the square, triangles BQE and BVC are similar, and we have

$$\frac{a}{VC} = \frac{BE}{BC} = BE.$$

Letting $VC = n$ and $UB = m$, we have

$$BE = \frac{a}{n},$$

and similarly

$$EC = \frac{a}{m}.$$

Therefore $BE + EC = BC$ yields

$$a\left(\frac{1}{n} + \frac{1}{m}\right) = 1, \quad a\left(\frac{m+n}{mn}\right) = 1, \quad \text{and} \quad a = \frac{mn}{m+n}.$$

Similarly, from Figure 3 we have that

$$b = \frac{(1-m)(1-n)}{(1-m)+(1-n)},$$

giving

$$\begin{aligned} 2 \cdot PUQV & \\ &= a + b = \frac{mn}{m+n} + \frac{(1-m)(1-n)}{(1-m)+(1-n)} \\ &= \frac{2mn - m^2n - mn^2 + m - m^2 - mn + m^2n + n - mn - n^2 + mn^2}{(m+n)(2-m-n)} \\ &= \frac{m+n-m^2-n^2}{(m+n)(2-m-n)}. \end{aligned}$$

Now, if $m = n$, then UV would be parallel to BC and thus divide the square into two rectangles with centers P and Q (Figure 4). In this case,

$$a = \frac{n}{2} \quad \text{and} \quad b = \frac{1-n}{2},$$

giving

$$a + b = \frac{1}{2}.$$

It is a small step to the conjecture that such a symmetrical situation provides the maximum area of $PUQV$. Accordingly, let us try to show for all m

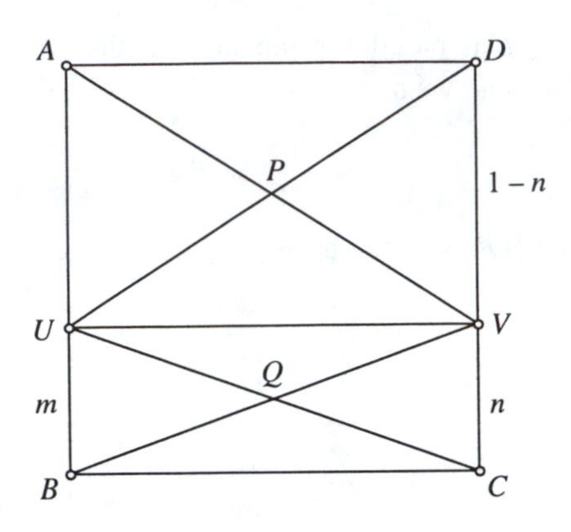

FIGURE 4

and n that $a+b$ never exceeds $\frac{1}{2}$:

$$a+b=\frac{m+n-m^2-n^2}{(m+n)(2-m-n)}\leq\frac{1}{2}.$$

This is equivalent to

$$2m+2n-2m^2-2n^2\leq 2m-m^2-mn+2n-mn-n^2,$$

and to

$$0\leq m^2-2mn+n^2$$
$$0\leq(m-n)^2,$$

which is obviously true, with equality if and only if $m=n$. Therefore

$$\max PUQV=\max\left(\frac{a+b}{2}\right)=\frac{1}{4},$$

and the maximum occurs whenever UV is parallel to BC and only in such cases.

SECTION 26

A Function of Exponential Order

Consider the product

$$f_n(x) = (1 - x^{a_1})(1 - x^{a_2})(1 - x^{a_3}) \cdots (1 - x^{a_n}),$$

where the exponents are positive integers. When $f_n(x)$ is multiplied out, a total of 2^n terms are obtained before simplification, each bearing a coefficient of $+1$ or -1. In the simplification, some terms might cancel and others accumulate sizeable coefficients. For example, in the case of

$$(1 - x)(1 - x^2)(1 - x^3)(1 - x^4) = 1 - x - x^2 + 2x^5 - x^8 - x^9 + x^{10},$$

fully half of the 16 original terms cancel. In terms of the notation

$$\begin{aligned} f_n(x) &= (1 - x^{a_1})(1 - x^{a_2})(1 - x^{a_3}) \cdots (1 - x^{a_n}) \\ &= 1 + b_1 x + b_2 x^2 + \cdots + b_k x^k, \qquad (k = a_1 + a_2 + \cdots + a_n), \end{aligned}$$

the number of terms which do **not** cancel is

$$N_n = 1 + |b_1| + |b_2| + \cdots + |b_k|.$$

Clearly N_n depends upon the exponents $a_1, a_2, \ldots, a_n$. Now, rather than keeping n fixed and considering the changes in N_n as the a_i are varied, let us settle on an infinite sequence of exponents $\{a_i\}$ and consider how N_n changes as n runs through the values $1, 2, 3, \ldots$. Relative to the chosen sequence, then, N_n is just a function of n.

An exact formula for N_n is a lot to ask. Indeed, we shall be content to discover whether the function for N_n is exponential, that is, with n appearing in an exponent, such as a^n $(a > 1)$, or is merely a polynomial in n, $c_r n^r + c_{r-1} n^{r-1} + \cdots + c_1 n + c_0$. Exponential functions are of a higher order than polynomials in the sense that, no matter how little the base might exceed 1, the values of a^n eventually overtake and overwhelm those of any polynomial, however great its degree or coefficients. For example, for $n = 1, 2, \ldots, 95$, the polynomial $f(n) = n^2$ provides greater values than the exponential function $g(n) = \left(\frac{11}{10}\right)^n$, but for $n \geq 96$, $g(n)$ pulls away rapidly:

$$\begin{aligned} f(95) &= 9025, & g(95) &= 8556.67\ldots \\ f(96) &= 9216, & g(96) &= 9412.34\ldots \\ f(100) &= 10000, & g(100) &= 13780.6\ldots \\ f(125) &= 15625, & g(125) &= 149308.88\ldots \\ &\ldots\ldots\ldots\ldots\ldots\ldots\ldots\ldots\ldots\ldots \end{aligned}$$

Bigger degrees and coefficients in a polynomial merely prolong the chase.

We will show that, for a broad class of sequences $\{a_i\}$, the value of N_n does indeed grow exponentially with n. Our result is due to Laszlo Szèkely (Hungary) and Bruce Richmond (Waterloo), whose interest in the subject arose in connection with a problem by Adrian Lewis (Waterloo) and Peter Borwein (Dalhousie). It is a pleasure to thank Bruce for showing me this neat piece of work.

The Result

N_n grows exponentially with n for all sequences $\{a_i\}$ for which there exists a prime number p which does **not** divide any of the integers a_i. In fact,

$$N_n \geq [p^{1/(p-1)}]^n \quad \text{for any such prime } p.$$

Clearly there are many sequences which do contain every prime p among the divisors of its terms; the obvious example is $\{1, 2, 3, \ldots\}$. However, the primality constraint still leaves a significant collection of sequences to which the result applies.

The Proof

Our task, then, is to get a line on the value of

$$N_n = 1 + |b_1| + |b_2| + \cdots + |b_k|.$$

Mercifully, we seek only a lower bound for it.

Clearly,

$$\begin{aligned} |f_n(x)| &= |1 + b_1x + b_2x^2 + \cdots + b_kx^k| \\ &\leq 1 + |b_1x| + |b_2x^2| + \cdots + |b_kx^k| \\ &= 1 + |b_1||x| + |b_2||x^2| + \cdots + |b_k||x^k|. \end{aligned}$$

In order to extract the number N_n from this, we need to substitute for x a value which makes every factor $|x^r|$ equal to 1. Accordingly, for $x = q = e^{2\pi i/p}$, the fundamental non-real pth-root of unity, where p is the distinguished prime which does not divide any a_i, we have

$$|b_i q^t| = |b_i||q^t| = |b_i||q|^t = |b_i| \cdot 1 = |b_i|.$$

Therefore, for each positive integer t we have

$$\begin{aligned} |f_n(q^t)| &= |1 + b_1 q^t + b_2 q^{2t} + \cdots + b_k q^{kt}| \\ &\leq 1 + |b_1 q^t| + |b_2 q^{2t}| + \cdots + |b_k q^{kt}| \\ &= 1 + |b_1| + |b_2| + \cdots + |b_k| \\ &= N_n. \end{aligned}$$

Adding the $p-1$ such inequalitites for $t = 1, 2, \ldots, p-1$, we obtain

$$|f_n(q)| + |f_n(q^2)| + |f_n(q^3)| + \cdots + |f_n(q^{p-1})| \leq (p-1)N_n,$$

giving

$$Nn \geq \frac{1}{p-1}[|f_n(q)| + |f_n(q^2)| + |f_n(q^3)| + \cdots + |f_n(q^{p-1})|].$$

That is to say, N_n is at least as great as the arithmetic mean A of the $p-1$ positive numbers $|f_n(q)|, |f_n(q^2)|, \ldots, |f_n(q^{p-1})|$.

From the initial factored form of the function $f_n(x)$, we have

$$\begin{aligned} |f_n(q)| &= |(1-q^{a_1})(1-q^{a_2})\cdots(1-q^{a_n})|, \\ |f_n(q^2)| &= |(1-q^{2a_1})(1-q^{2a_2})\cdots(1-q^{2a_n})|, \\ &\vdots \\ |f_n(q^{p-1})| &= |(1-q^{(p-1)a_1})(1-q^{(p-1)a_2})\cdots(1-q^{(p-1)a_n})|. \end{aligned}$$

By the arithmetic mean-geometric mean inequality, then, we have

$$N_n \geq A \geq G,$$

and, calculating G by taking the factors in columns, we get

$$N_n \geq G = \left[\prod_{k=1}^{n} |(1-q^{a_k})(1-q^{2a_k})\cdots(1-q^{(p-1)a_k})|\right]^{1/(p-1)}.$$

Now, $q^{a_k} = e^{2\pi i a_k/p}$, where p does not divide a_k. Thus q^{a_k} is one of the non-real pth-roots of unity, and since p is a prime number, its $p-1$ powers

$$q^{a_k}, q^{2a_k}, \ldots, q^{(p-1)a_k}$$

constitute the entire set of $p-1$ non-real pth-roots of unity. With the real root 1, these give all p of the pth-roots of unity, and we have the factoring

$$x^p - 1 = (x-1)(x-q^{a_k})(x-q^{2a_k})\cdots(x-q^{(p-1)a_k}).$$

Hence, for $x \neq 1$,

$$(x-q^{a_k})(x-q^{2a_k})\cdots(x-q^{(p-1)a_k}) = \frac{x^p-1}{x-1}.$$

As observed, this equation is not valid for $x = 1$. However, because the function on the left side is **continuous**, its value at $x = 1$ is equal to the limit of its values as $x \to 1$, and although the right side is discontinuous at $x = 1$, its **limit** as $x \to 1$ is the **same limit** that is approached by the left side. Hence

$$\begin{aligned}&(1-q^{a_k})(1-q^{2a_k})\cdots(1-q^{(p-1)a_k})\\ &\quad= \lim_{x\to 1}\frac{x^p-1}{x-1} = \lim_{x\to 1}\frac{px^{p-1}}{1}\ \text{(L'Hospital)} = p\end{aligned}$$

Thus we have

$$N_n \geq G = \left[\prod_{k=1}^{n} p\right]^{1/(p-1)} = [p^n]^{1/(p-1)} = \left[p^{1/(p-1)}\right]^n,$$

and since $p^{1/(p-1)} > 1$, then N_n grows exponentially, completing the proof.

Exercises

Here are a few exercises for your amusement.

1. (From a Year 8 Bulgarian Competition)

   From the midpoint D of the base AB of isosceles triangle ABC, DE is drawn perpendicular to BC (Figure 1). If H is the midpoint of DE, prove that AE and CH are perpendicular.

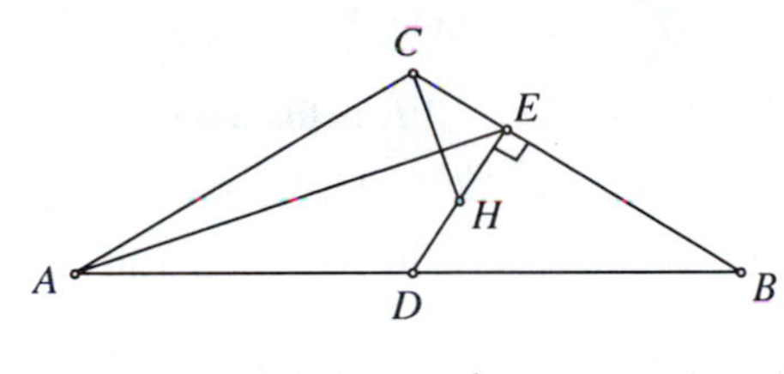

FIGURE 1

2. (From the 1990 Tournament of the Towns—Years 8, 9, 10)

   P is the midpoint, and M any other point, of the minor arc AC of the circumcircle of equilateral triangle ABC. N is the midpoint of BM and K

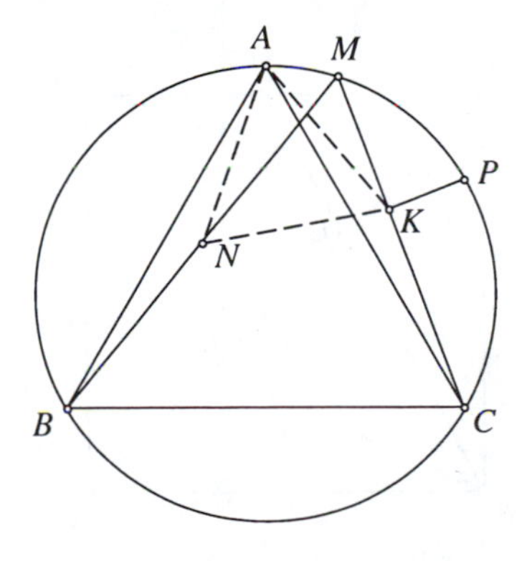

FIGURE 2

is the foot of the perpendicular from P to MC. Prove $\triangle ANK$ is equilateral for all choices of M.

3. (From the 1896 Eötvös Competition (Hungary))

   Construct an acute angled triangle ABC given the feet D, E, F of its altitudes.

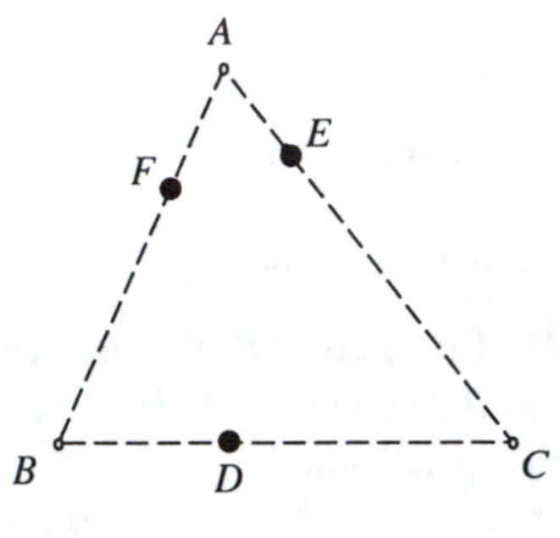

FIGURE 3

4. The sequence $A = \{a_1, a_2, a_3, \ldots\}$ is defined by

   $$a_n = 1 + 2^2 + 3^3 + \cdots + n^n.$$

   Prove that A contains infinitely many odd composite terms.

5. (A gem from Bulgaria)

   E and F are variable points on the sides AD and CD, respectively, of square $ABCD$ such that $DE = CF$ (Figure 4). BE and BF divide AC into three segments AG, GH, and HC which, as it turns out, can always be arranged to form a triangle. Prove that every such triangle has an angle of 60°.

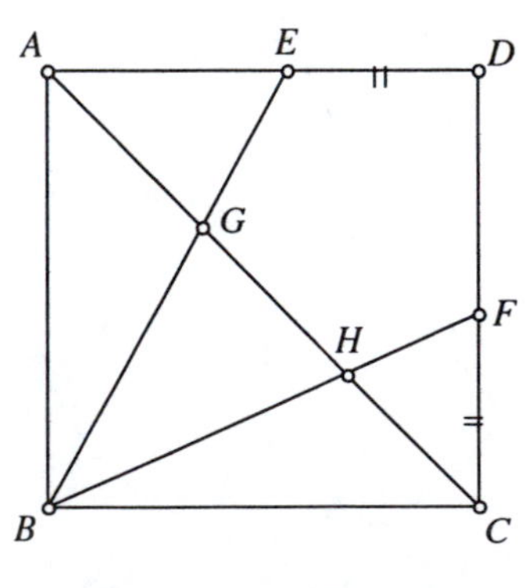

FIGURE 4

6. Find all solutions in positive integers (a, b, c) of the equation

$$a^b + b^c = abc.$$

7. (From an East German Olympiad)

   Determine all the **integers** in the sequence $\{a_n\}$ defined by

$$a_1 = 1, \quad a_{n+1} = 2a_n + \sqrt{3a_n^2 + 1}.$$

8. (Due to Gordon Lessells, Limerick University, Ireland)

   The real number x satisfies all of the inequalities

$$2^k < x^k + x^{k+1} < 2^{k+1}$$

   for $k = 1, 2, \ldots, n$. What is the greatest possible value of n?

9. (From the 1992 U.S.A. Olympiad)

   For $n = 0, 1, 2, \ldots$, let the product $p(n)$ be defined by

$$p(n) = 9(99)(9999)\cdots(10^{2^n} - 1),$$

   where the number of 9's doubles with each factor. Determine a formula for the sum $s(n)$ of the digits in the result when $p(n)$ is multiplied out.

10. In isosceles $\triangle ABC$, the length of an altitude to one of the equal sides AB and BC is four units. A straight line through B crosses the base AC at D so that the incircle of $\triangle ABD$ is the same size as the excircle of $\triangle BDC$ which is opposite B (Figure 5). Prove the circles are unit circles.

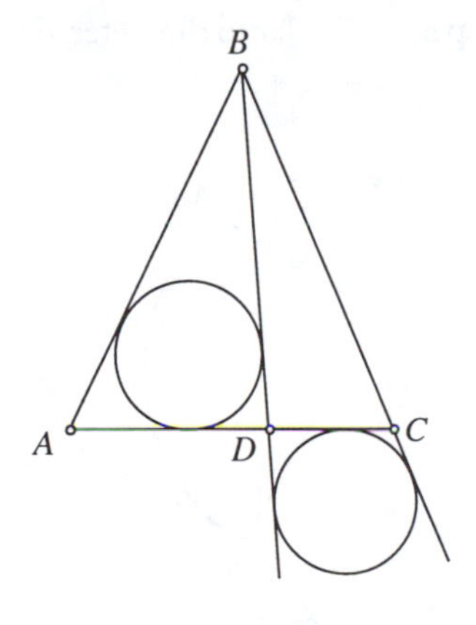

FIGURE 5

11. The tangent from an external point P to a circle, center O, meets the circle at A. The perpendicular from A meets OP at B and the midpoint of BP is M. If the tangent to the circle from M meets the circle at C, prove that $\angle BCP$ is a right angle.

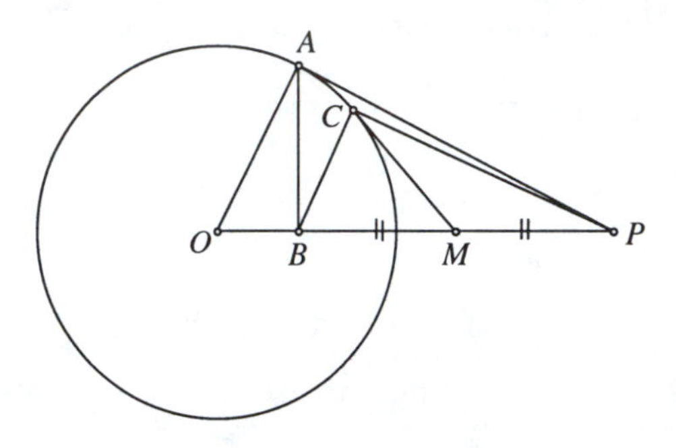

FIGURE 6

12. What masses would you suspend at the vertices of an acute-angled triangle in order to determine a system whose center of mass occurs at the orthocenter of the triangle ?

13. If $a_1, a_2, \ldots, a_n$ are distinct positive integers none of which is divisible by a prime number bigger than 3, prove for all $n = 1, 2, 3, \ldots$ that

$$\frac{1}{a_1} + \frac{1}{a_2} + \cdots + \frac{1}{a_n} < 3.$$

14. The tangents from an external point P touch a circle, center O, at A and B (Figure 7). AD is perpendicular to diameter BC. Prove that CP bisects AD.

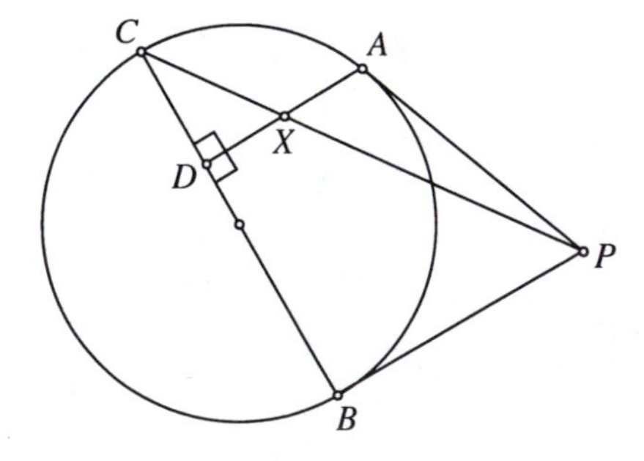

FIGURE 7

15. (From an article by L. J. Paradiso in the *American Mathematical Monthly*, 1929, 87–89.)

$$\text{Rationalize } \frac{1}{2\sqrt[3]{2} - 5\sqrt[3]{5}}.$$

16. Given a parabola, just the curve, determine a straightedge and compass construction for its focus.

17. If $\{t_1, t_2, t_3, \ldots\}$ is a sequence of positive real numbers such that
    (i) $t_n \cdot t_{n+1} = n$ for all $n \geq 1$, and
    (ii) $\frac{t_n}{t_{n+1}} \to 1$ as $n \to \infty$,

    prove that the only possible value for the first term is $\sqrt{\frac{2}{\pi}}$.

18. (Problem 870, *Pi Mu Epsilon Journal*, Fall, 1995, 228: proposed by Grattan P. Murphy, University of Maine, Orono, Maine.)

    Prove that some digit in the decimal representation of $N = (7^7)^7 \cdot 7^7 \cdot 77$ occurs at least six times.

19. (Note 2548, by F. M. Arscott, 1955, 232)

    There are well known tests for divisibility by 9 and 11 and various other special cases. Prove that the following is a valid test for divisibility by 19.

    To test a positive integer n, remove its last digit d and add $2d$ to the number which is left, and keep doing this until a number r is reached which can easily be checked mentally for divisibility by 19. Then n is divisible by 19 if and only if r is.

    For example, for $n = 704836$, we proceed to 70495, 7059, 723, 78, 23, which is not divisible by 19; hence neither is 704836. (This test is due to J. Kashangaki, Makerere College, University of East Africa.)

20. (Problem 66.A, 1987, 237)

    If n positive real numbers have sum s, what is the least possible value of the sum of their reciprocals?

21. (Due to the ingenious Victor Thébault; 1937, 52)

    E and F are chosen arbitrarily on the sides AC and AB, respectively, of $\triangle ABC$ (Figure 8). The circumcircles of triangles ABE and ACF meet

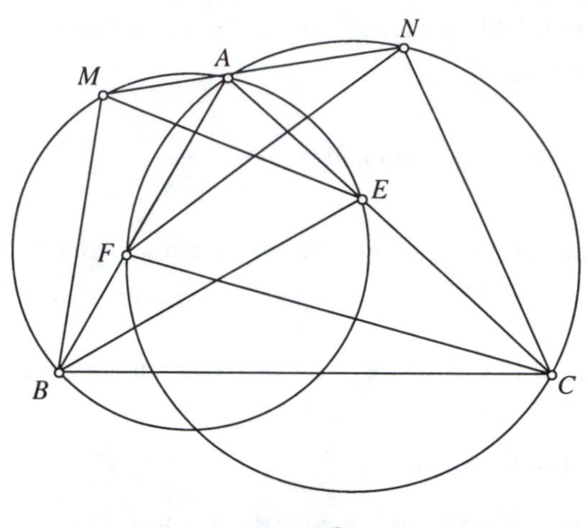

FIGURE 8

the bisector of the exterior angle at A in M and N. Prove that triangles MBE and NCF are isosceles and similar.

Solutions to the Exercises

1. Let EF and HJ be perpendiculars from E and H to AB and CD, respectively (Figure 1). Let $\angle B = x$ and $y = 90^\circ - x$; then clearly, angles x and y occur in the right angled triangles as shown in the figure. Finally, let rectangular axes be assigned with the origin at A and the positive x-axis along AB.

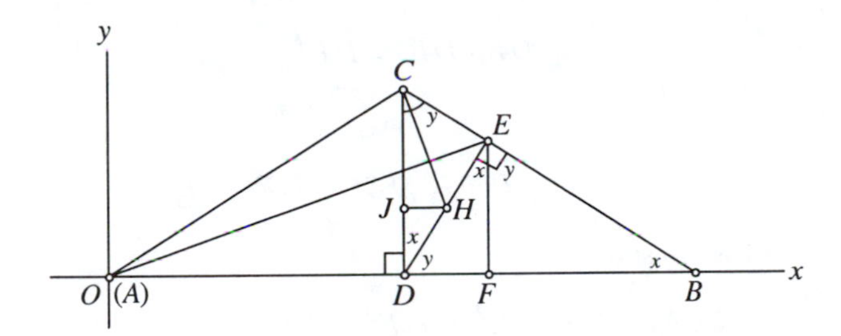

FIGURE 1

Then

$$\text{slope } AE = \frac{EF}{AF} \quad \text{and} \quad \text{slope } CH = -\frac{CJ}{JH} = -\frac{CD - JD}{JH}.$$

Now, since H is the midpoint of DE, then, by similar triangles JDH and DEF,

$$JH = \frac{1}{2}DF \quad \text{and} \quad JD = \frac{1}{2}EF,$$

giving

$$\text{slope } CH = -\frac{CD - \frac{1}{2}EF}{\frac{1}{2}DF} = -\frac{2CD - EF}{DF}.$$

Thus we would like to show that

$$(\text{slope } AE) \cdot (\text{slope } CH) = -1,$$

i.e., that

$$\frac{EF}{AF} \cdot \frac{2CD - EF}{DF} = 1,$$
$$2EF \cdot CD - EF^2 = AF \cdot DF,$$

or

$$2EF \cdot CD = AF \cdot DF + EF^2.$$

Now,

$$AF = AD + DF = DB + DF,$$

giving

$$AF \cdot DF = DF \cdot DB + DF^2.$$

But, from right angles triangles,

$$DF \cdot DB = DE^2,$$

making

$$AF \cdot DF = DE^2 + DF^2.$$

Hence we wish to show that

$$\begin{aligned} 2EF \cdot CD &= DE^2 + DF^2 + EF^2 \\ &= DE^2 + DE^2 \qquad \text{(from } \triangle DEF\text{)} \end{aligned}$$

i.e., that

$$EF \cdot CD = DE^2.$$

But,

$$\sin y = \frac{EF}{DE} = \frac{DE}{CD} \qquad \text{(from triangles } DEF \text{ and } CDE\text{)},$$

from which the desired result follows directly.

2. If we could prove that triangles ANB and AKC are congruent, then $\triangle BAN$ would equal $\triangle CAK$, making $\angle NAK = \angle BAC = 60°$ (Figure 2). Also, AN would equal AK, making $\triangle ANK$ an isosceles triangle with a $60°$ angle between its equal sides, from which the conclusion follows.

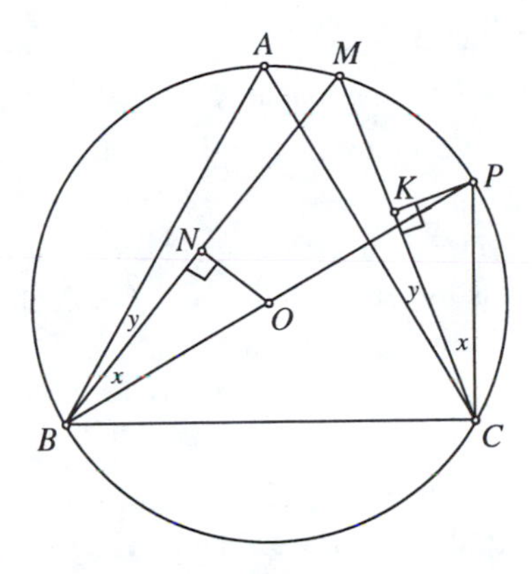

FIGURE 2

Toward this goal it is clear that $AB = AC$ and $\angle ABN = \angle ACK$ (on the arc AM). It remains only to show that $BN = CK$. But, as we shall see, this follows from the congruency of triangles BON and CPK, where O is the center of the circle. Since the segment to the midpoint of a chord from the center of a circle is perpendicular to the chord, these triangles are both right angled. Also $\angle NBO = \angle KCP$ on arc MP. Finally, OB equals the radius of the circle, and so does CP because it subtends a $30°$ angle at B (PB bisects the $60°$ angle at B, making $\angle PBC = 30°$, which in turn makes $\angle POC = 2\angle PBC = 60°$, implying ΔPOC is equilateral with side equal to the radius).

3. Let the orthocenter of $\angle ABC$ be H. The right angles at D, E, F imply several cyclic quadrilaterals (Figure 3):

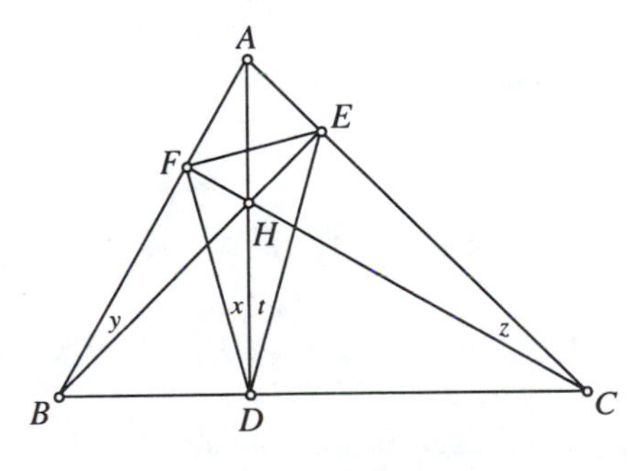

FIGURE 3

from $FBDH$, we have equal angles x and y on chord FH;
from $FBCE$, we have equal angles y and z on chord EF; and
from $HDCE$, we have equal angles z and t on chord EH.

Hence $x = t$ and DH is the bisector of angle D in $\triangle DEF$. Similarly HE and HF bisects the other angles in $\triangle DEF$; that is to say, the angle bisectors of $\angle DEF$ are the altitudes of $\triangle ABC$. Thus the sides of $\triangle ABC$ are obtained by drawing perpendiculars to the angle bisectors of $\triangle DEF$ at D, E, F.

4. Clearly n^n is odd or even as n is odd or even. Thus A begins with two odd terms and thereafter alternates two even terms and two odd terms:

$$A = \{1, 5, 32, 288, 3413, \ldots, \text{odd, odd, even, even}, \ldots\}.$$

Therefore A contains an infinite number of odd terms. Now, modulo 3, the respective integers

$$\begin{aligned} &1, 2^2, 3^3, 4^4, 5^5, 6^6, 7^7, 8^8, 9^9, 10^{10}, 11^{11}, 12^{12}, 13^{13}, 14^{14}, 15^{15}, 16^{16}, 17^{17}, 18^{18}, 19^{19}, \ldots \\ \equiv\ &1, 1, 0, 1, 2, 0, 1, 1, 0, 1, 2, 0, 1, 1, 0, 1, 2, 0, 1, \ldots, \end{aligned}$$

a sequence with period 6. Hence, adding these residues, we find (mod 3) that

$$a_n \equiv 1, 2, 2, 0, 2, 2, 0, 1, 1, 2, 1, 1, 2, 0, 0, 1, 0, 0, 1, 2, 2, 0, \ldots,$$

a sequence with period 18.

Therefore every eighteenth term of A is divisible by 3 and, being bigger than 3, is composite. Of course many such terms are even numbers. However, since the terms of A alternate "odd, odd, even, even," if a_{18k} is even, then a_{18k+18} is odd, and vice versa, and so every 36th term of A, starting with a_{18} is an odd composite number, and the conclusion follows.

5. If the lengths of the sides of a triangle are x, y, and z, then, by the law of cosines, the angle opposite the side of length x is 60° if and only if

$$x^2 = y^2 + z^2 - 2yz\cos 60^\circ = y^2 + z^2 - yz.$$

Thus we need to show such a relation between the lengths of AG, GH, and HC.

Also, if one angle of a triangle is 60°, either the other two angles are each 60° or one is bigger than 60° and the other smaller. Thus, if a triangle is not equilateral, a 60° angle is opposite the side of intermediate length (a bigger angle is opposite a longer side). Clearly, when F is near C, HC is the smallest side and one of AG or GH is the side of intermediate length. On the other hand, when E is near A, AG is the smallest side, and so it is not inconceivable that it is GH which is always the side of intermediate length. Therefore, let us direct our efforts toward establishing

$$GH^2 = AG^2 + HC^2 - AG \cdot HC.$$

When a problem is posed in a Euclidean context, one is inclined to pursue a Euclidean solution. However, let us avail ourselves of Descartes's wonderful notions and proceed analytically.

Accordingly, suppose $ABCD$ is the unit square and let rectangular axes be imposed with the origin at B and positive x-axis along BC (Figure 4). Also, let $DE = CF = a$, making $F(1, a)$ and $E(1 - a, 1)$.

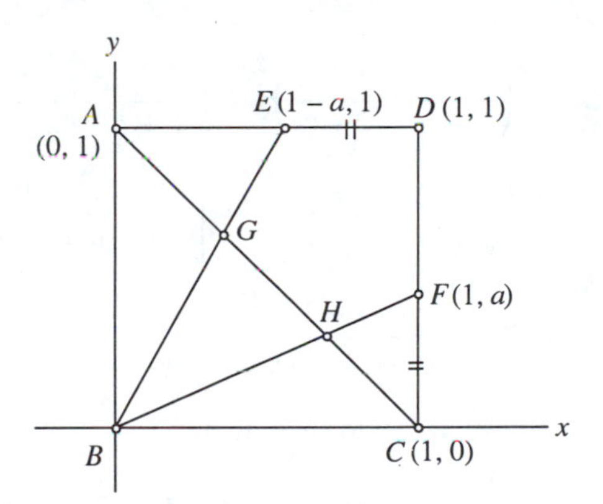

FIGURE 4

Then the equations of AC and BF are simply

$$x + y = 1 \quad \text{and} \quad y = ax,$$

and solving for H, we get

$$x + ax = 1,$$

making

$$H\left(\frac{1}{1+a}, \frac{a}{1+a}\right).$$

Similarly, the equation of BE is $y = \frac{x}{1-a}$, and solving for G, we have

$$x + \frac{x}{1-a} = 1, \quad x\left(\frac{1-a+1}{1-a}\right) = 1, \quad x\left(\frac{2-a}{1-a}\right) = 1,$$

making

$$G\left(\frac{1-a}{2-a}, \frac{1}{2-a}\right).$$

Hence

$$AG^2 = \left(\frac{1-a}{2-a}\right)^2 + \left(1 - \frac{1}{2-a}\right)^2 = 2\left(\frac{1-a}{2-a}\right)^2;$$

$$HC^2 = \left(1 - \frac{1}{1+a}\right)^2 + \left(\frac{a}{1+a}\right)^2 = 2\left(\frac{a}{1+a}\right)^2;$$

and

$$\begin{aligned} GH^2 &= \left(\frac{1}{1+a} - \frac{1-a}{2-a}\right)^2 + \left(\frac{a}{1+a} - \frac{1}{2-a}\right)^2 \\ &= \left[\frac{2-a-1+a^2}{(1+a)(2-a)}\right]^2 + \left[\frac{2a-a^2-1-a}{(1+a)(2-a)}\right]^2 \\ &= 2\left[\frac{a^2-a+1}{(1+a)(2-a)}\right]^2. \end{aligned}$$

Therefore we would like to show that

$$2\left[\frac{a^2-a+1}{(1+a)(2-a)}\right]^2 = 2\left(\frac{1-a}{2-a}\right)^2 + 2\left(\frac{a}{1+a}\right)^2 - 2\left(\frac{1-a}{2-a}\right)\left(\frac{a}{1+a}\right)$$

that is,

$$\begin{aligned} (a2-a+1)^2 &= (1-a)^2(1+a)^2 + a^2(2-a)^2 \\ &\quad - a(1-a)(1+a)(2-a), \\ a^4 + a^2 + 1 - 2a^3 + 2a^2 - 2a &= (1-a^2)^2 + a^2(4-4a+a^2) \\ &\quad - (1-a^2)(2a-a^2), \\ a^4 - 2a^3 + 3a^2 - 2a + 1 &= 1 - 2a^2 + a^4 + 4a^2 - 4a^3 + a^4 \\ &\quad - 2a + a^2 + 2a^3 - a^4 \\ &= a^4 - 2a^3 + 3a^2 - 2a + 1, \end{aligned}$$

which is so. Since all these operations are reversible, the conclusion follows.

6. While there do not seem to be many promising ways to begin a solution, we might observe that b and c occur as exponents on the left side and a does not; moreover, b occurs there twice and is therefore likely to be the dominant variable in the equation. Also, since powers on the left are weighed against simple factors on the right, surely the left side will overwhelm the right side for any sizeable values of the variables. Accordingly, let us investigate the solutions for $b = 1, 2, 3, \ldots$, hoping that we won't have to go very far.

   (i) $b = 1$. In this case, we seek a and c such that

$$a + 1 = ac,$$
$$1 = a(c - 1).$$

   Being positive factors of 1, this requires $a = c - 1 = 1$, and we have the solution $(a, b, c) = (1, 1, 2)$.

   (ii) $b = 2$. This leaves us to solve

$$a^2 + 2^c = 2ac, \quad \text{that is,} \quad a^2 - 2ac + 2^c = 0.$$

   Since c is an exponent, let us consider solutions for $c = 1, 2, 3, \ldots$. For $c = 1$, we have

$$a^2 - 2a + 2 = 0,$$
$$(a - 1)^2 + 1 = 0,$$

   which has no integer solution. For $c = 2$, we have

$$a^2 - 4a + 4 = 0,$$

   giving $a = 2$ and the solution $(2, 2, 2)$. For $c = 3$, we have

$$a^2 - 6a + 8 = (a - 2)(a - 4) = 0,$$

   yielding

$$a = 2 \text{ and } 4,$$

   and the solutions

$$(2, 2, 3) \text{ and } (4, 2, 3).$$

   For $c = 4$, we get

$$a^2 - 8a + 16 = 0,$$

   giving $a = 4$ and the solution $(4, 2, 4)$.

It turns out that there are no solutions for $c \geq 5$, for at this point

$$2^c > c^2,$$

which is easily established by induction as follows: for $c = 5$, we have

$$2^c = 2^5 = 32 > 5^2 = c^2,$$

and if $2^c > c^2$ holds for some $c \geq 5$, then

$$\begin{aligned} 2^{c+1} &= 2 \cdot 2^c > 2c^2 \\ &= (c+1)^2 + c(c-2) - 1 \geq (c+1)^2 + 14 > (c+1)^2. \end{aligned}$$

Because of this, we have that

$$a^2 + 2^c > a^2 + c^2 \geq 2ac, \text{ (by the A.M.–G.M inequality)},$$

denying the equality $a^2 + 2^c = 2ac$.

(iii) $b \geq 3$. With $a = 1$, the equation reduces to

$$1 + b^c = bc.$$

Now, it is easily proved by induction (the details are omitted) that

$$b^c \geq bc,$$

implying

$$1 + b^c > bc.$$

Hence there is no solution for $a = 1$.

Finally, consider $a = 2$. Since $b \geq 3$, then

$$3 \cdot 2^{b-3} \geq b \quad \text{(by induction)},$$

giving

$$2^{b-2} \geq \frac{2}{3}b.$$

Since $a \geq 2$, then

$$a^{b-2} \geq 2^{b-2} \geq \frac{2}{3}b,$$

and we have

$$a^b \geq \frac{2}{3}a^2b.$$

Similarly, induction yields

$$3^c \geq 2c^2,$$

giving

$$3^{c-1} \geq \frac{2}{3}c^2,$$

and with $b \geq 3$, we have

$$b^{c-1} \geq 3^{c-1} \geq \frac{2}{3}c^2,$$

from which

$$b^c \geq \frac{2}{3}bc^2.$$

Thus

$$\begin{aligned} a^b + b^c &\geq \frac{2}{3}a^2b + \frac{2}{3}bc^2 \\ &\geq 2\sqrt{\frac{2}{3}a^2b \cdot \frac{2}{3}bc^2} \qquad \text{(by the A.M.–G.M. inequality)} \\ &= \frac{4}{3}abc \\ &> abc, \end{aligned}$$

showing there are no solutions for $a \geq 2$.

Thus the five solutions are

$$(1, 1, 2), (2, 2, 2), (4, 2, 3), (2, 2, 3), \text{ and } (4, 2, 4).$$

7. Calculating a few terms, we get

$$\{a_n\} = \{1, 4, 15, 56, 209, 780, \ldots\},$$

suggesting that a_n might always be an integer. Since the radical in the definition is so awkward, perhaps there is an alternative definition that would make the integer characteristics of a_n more evident. From our small sample, it appears that the terms might be related by the recursion

$$a_{n+1} = 4a_n - a_{n-1}.$$

If this is so, then it is certainly true that all a_n are integers. Hence let us solve this recursion to see whether its solutions $\{b_n\}$ is the same sequence as $\{a_n\}$.

Solving the auxiliary equation

$$x^2 - 4x + 1 = 0,$$

we get

$$x = \frac{4 \pm \sqrt{12}}{2} = 2 \pm \sqrt{3},$$

giving the solution

$$b_n = a(2+\sqrt{3})^n + b(2-\sqrt{3})^n$$

for some constants a and b. Observing that $a_0 = 0$ is in keeping with the definition of a_n ($a_0 = 0$ implies $a_1 = 2 \cdot 0 + \sqrt{3 \cdot 0 + 1} = 1$), we can simplify the calculation of a and b by using the cases $n = 0$ and 1 instead of $n = 1$ and 2. Thus, with the initial values $b_0 = 0$ and $b_1 = 1$, we have

$$b_0 = a + b = 0, \quad \text{giving} \quad b = -a$$

and

$$b_n = a[(2+\sqrt{3})^n - (2-\sqrt{3})^n];$$

then

$$b_1 = a(2\sqrt{3}) = 1, \quad \text{and} \quad a = \frac{1}{2\sqrt{3}}.$$

Letting $2 + \sqrt{3} = \alpha$ and $2 - \sqrt{3} = \beta$, we have $\alpha - \beta = 2\sqrt{3}$ and the formula

$$b_n = \frac{\alpha^n - \beta^n}{\alpha - \beta}.$$

We note also that

$$\alpha + \beta = 4 \quad \text{and} \quad \alpha\beta = 1.$$

It would please us to be able to show that

$$b_{n+1} = 2b_n + \sqrt{3b_n^2 + 1},$$

that is, that

$$(b_{n+1} - 2b_n)^2 = 3b_n^2 + 1.$$

We have

$$(b_{n+1}-2b_n)^2 = \left[\frac{\alpha^{n+1}-\beta^{n+1}}{\alpha-\beta} - 2\cdot\frac{\alpha^n-\beta^n}{\alpha-\beta}\right]^2$$
$$= \left(\frac{1}{\alpha-\beta}\right)^2 \left[\alpha^n(\alpha-2)-\beta^n(\beta-2)\right]^2.$$

Recalling that $\alpha-\beta=2\sqrt{3}$, and noting that $\alpha-2=\sqrt{3}$ and $\beta-2=-\sqrt{3}$, we have

$$(b_{n+1}-2b_n)^2 = \frac{1}{12}[\sqrt{3}(\alpha^n+\beta^n)]^2 = \frac{1}{4}(\alpha^n+\beta^n)^2.$$

Also

$$3b_n^2+1 = 3\left(\frac{\alpha^n-\beta^n}{\alpha-\beta}\right)^2+1 = \frac{3}{12}(\alpha^n-\beta^n)^2+1$$
$$= \frac{1}{4}[(\alpha^n-\beta^n)^2+4]$$
$$= \frac{1}{4}[(\alpha^n-\beta^n)^2+4\alpha^n\beta^n] \quad \text{(recall } \alpha\beta=1)$$
$$= \frac{1}{4}(\alpha^n+\beta^n)^2,$$

as hoped. Thus $\{b_n\}$ and $\{a_n\}$ are the same sequence and it follows that all a_n are integers.

8. The first inequality restricts x to the two ranges determined by

$$2 < x+x^2 \quad \text{and} \quad x+x^2 < 4.$$

Now, $2 < x+x^2$ is satisfied by all real numbers **except** those lying in the closed interval between the roots -2 and 1 of $x^2+x-2=0$, i.e., for $x<-2$ and $x>1$ (Figure 5). Similarly, $x+x^2<4$ is satisfied only by the values in the open interval between the roots of $x^2+x-4=0$, i.e., for

$$x \in \left(\frac{-1-\sqrt{17}}{2}, \frac{-1+\sqrt{17}}{2}\right).$$

Thus the first inequality confines x to either of the two open intervals

$$\frac{-1-\sqrt{17}}{2} < x < -2 \quad \text{and} \quad 1 < x < \frac{-1+\sqrt{17}}{2}.$$

Now, there are values slightly less than $\frac{-1+\sqrt{17}}{2}$, e.g., $x=1.55$, which satisfy the first three inequalities, but no x in $1 < x < \frac{-1+\sqrt{17}}{2}$

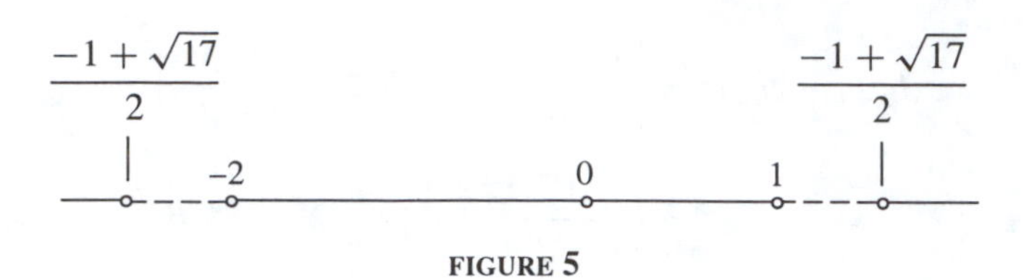

FIGURE 5

satisfies the fourth inequality: even for $x = \frac{1+\sqrt{17}}{2}$, the value of $x^4 + x^5 =$ 15.23 approximately, which is not big enough to be in the required range $(2^4, 2^5)$. For values in the interior of the interval, $x^4 + x^5$ is even smaller. Hence the maximum value of n is not more than 3.

Finally, consider the interval $\frac{-1-\sqrt{17}}{2} < x < -2$. We need only check whether any x in this interval satisfies the fourth inequality. But this is immediate, for with $x < -1$, the value of x^5 is negative and of greater magnitude than the positive number x^4, making $x^4 + x^5 < 0 < 2^4 = 16$. Thus no negative x can satisfy more than the first three inequalities and it follows that the maximum value of n is 3.

9. A glance at the first few cases immediately suggests the formula $s(n) = 9 \cdot 2^n$, which can be established nicely by an examination of the way $p(n)$ is derived from $p(n-1)$.

n	$p(n)$	$s(n)$
0	9	$9 = 9 \cdot 2^0$
1	891	$18 = 9 \cdot 2^1$
2	8909109	$36 = 9 \cdot 2^2$
3	890910891090891	$72 = 9 \cdot 2^3$

Clearly,

$$\begin{aligned} p(n) &= p(n-1) \cdot (10^{2^n} - 1) \\ &= p(n-1) \cdot 10^{2^n} - p(n-1); \end{aligned}$$

that is to say, attach 2^n 0's on the end of $p(n-1)$ and subtract $p(n-1)$ from the result. For example, in determining $p(3)$, we would have

$$\begin{array}{rr} p(2) \cdot 10^{2^3}: & 890910900000000 \\ p(2): & 8909109 \\ \hline p(3): & 890910891090891. \end{array}$$

As we shall see, the general procedure is as follows:

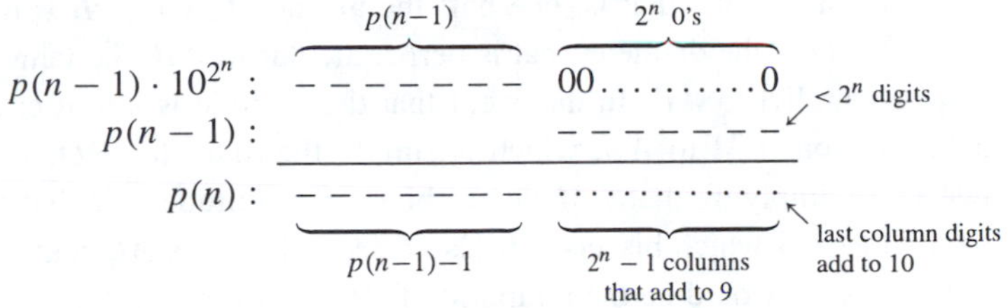

$$\begin{array}{rcc}
 & \overbrace{\qquad\qquad\qquad}^{p(n-1)} & \overbrace{\qquad\qquad\qquad}^{2^n \text{ 0's}} \\
p(n-1)\cdot 10^{2^n}: & \text{- - - - - - - -} & 00\ldots\ldots\ldots 0 \\
p(n-1): & & \text{- - - - - - - -} \quad \swarrow < 2^n \text{ digits} \\
\hline
p(n): & \text{- - - - - - - -} & \ldots\ldots\ldots\ldots \quad \nwarrow \\
 & \underbrace{\qquad\qquad\qquad}_{p(n-1)-1} & \underbrace{\qquad\qquad\qquad}_{\substack{2^n-1 \text{ columns} \\ \text{that add to 9}}}
\end{array}$$

last column digits add to 10

Because

$$\begin{aligned}
p(n-1) &= 9(99)(9999)\ldots(10^{2^{n-1}}-1) \\
&< 10(10^2)(10^4)\ldots(10^{2^{n-1}}) \\
&= 10^{1+2+4+\cdots 2^{n-1}} \\
&= 10^{2^n-1},
\end{aligned}$$

the number of digits in $p(n-1)$ is less than 2^n. Thus the only affect the subtraction has on the digits of $p(n-1)$ at the beginning of $p(n-1)\cdot 10^{2^n}$ is to reduce the final digit by 1; other than this, $p(n-1)$ carries over unchanged to the beginning of $p(n)$.

Now, subtracting 1 from a number can alter the sum of the digits drastically (e.g., $10000 - 1 = 9999$). However, since the last digit of $p(n)$ clearly alternates $9, 1, 9, 1, \ldots$, the reduction of $p(n-1)$ by 1 merely reduces the sum of the digits to $s(n-1) - 1$.

Another consequence of the final digit of $p(n-1)$ being 9 or 1 is that the first subtraction to be carried out, in the rightmost column, is either 9 from 10, resulting in 1, or 1 from 10, resulting in 9; thus this column always holds both a 9 and a 1 to give a sum of 10 for the digits in this column. Since the next $2^n - 1$ digits in the top line are 0's, there is a carryover in each case, with the result that the digits in these columns always add up to 9. Therefore, adding up the digits in the second and third lines in the above array gives

$$[s(n-1)-1] + 9\cdot(2^n-1) + 10.$$

But this is clearly just the sum of the digits in $p(n-1)$ and $p(n)$, and we have

$$s(n-1) + s(n) = [s(n-1)-1] + 9\cdot(2^n-1) + 10,$$

giving

$$s(n) = 9\cdot 2^n,$$

as desired.

10. We wish to show that the incircle of $\triangle ABD$ is a unit circle, implying a diameter of 2 units, that is, one-half the altitude CE to AB (Figure 6). Now, if PQ is the diameter that is perpendicular to AB, the tangent LM at Q is parallel to AB. In the event that the incircle **is** a unit circle, the distance from LM to AB, which is simply the diameter PQ, would be two units, implying that LM would bisect the altitude CE; but the line parallel to AB which bisects CE also bisects the sides AC and BC, and in this case L would be the midpoint of AC. Conversely, if we can show that the tangent at Q crosses AC at its midpoint, then the tangent would bisect CE and the diameter of the incircle would be 2 units, as desired. We need only show, then, that the point L where the tangent at Q crosses AC is in fact the midpoint of AC.

    Let the lengths of the tangents be $AR = x$, $BS = y$, $DR = z$, and $RL = u$ (Figure 6). Since the circles are the same size, the equal opposite angles at D imply that the four tangents from D to the two circles have the same length z. Now, since LM is parallel to AB, we have equal corresponding angles α at A and L, and since $AB = BC$, then $\angle ACB$ is also α. Thus the supplementary angles at L and C, namely $\angle RLQ$ and $\angle TCU$, are equal, and again since the circles are the same size, it follows

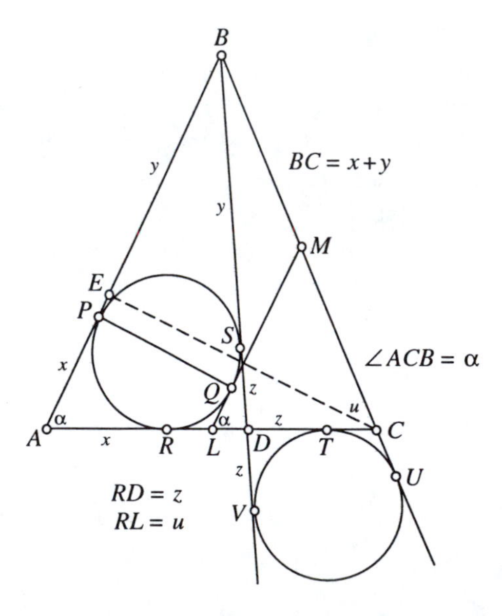

FIGURE 6

that the tangents to the respective circles from L and C are equal:

$$RL = TC = CU = u.$$

Hence

$$AL = x + RL = x + u,$$

and we would like to show that $AC = 2(x + u)$.

Now, since $AB = x + y = BC$ and $BS = y$, we have

$$\begin{aligned} x + u &= (x + y + u) - y \\ &= BU - y \\ &= BV - y \\ &= SV \\ &= 2z \\ &= RT \\ &= AC - (AR + TC) \\ &= AC - (x + u), \end{aligned}$$

giving the desired

$$2(x + u) = AC.$$

11. The problem reduces to proving $MB = MC$, for then the circle on diameter BP would go through C, making $\angle BCP$ a right angle (Figure 7).

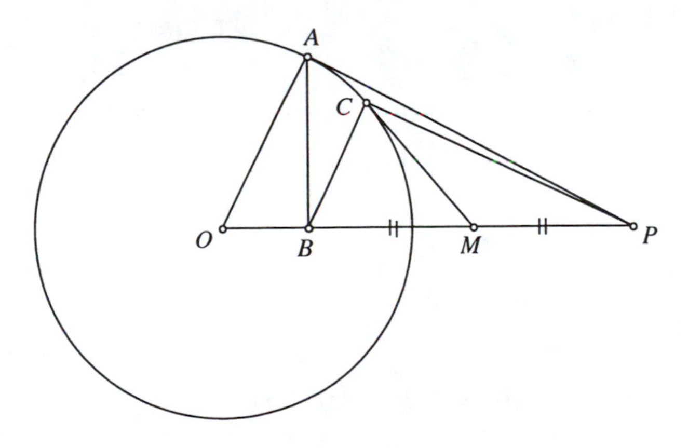

FIGURE 7

Let the radius of the given circle be 1. Now, AB is the altitude to the hypotenuse in right triangle OAP, and therefore a standard mean proportion yields

$$OA^2 = OP \cdot OB. \tag{1}$$

The Pythagorean theorem applied to triangles OAB and OCM (recall MC is a tangent) gives

$$AB^2 + OB^2 = 1, \tag{2}$$

and

$$MC^2 = OM^2 - OC^2 = OM^2 - 1. \tag{3}$$

Finally, it is clear that

$$OM = OB + BM \tag{4}$$

and

$$BP = OP - OB. \tag{5}$$

It remains only to manipulate these relations to see that $MB^2 = MC^2$, from which the desired $MB = MC$ completes the solution. From (1),

$$OA^2 = OP \cdot OB, \quad \text{i.e.,} \quad 1 = OP \cdot OB,$$

giving

$$OP = \frac{1}{OB}.$$

Then (5) gives

$$BP = OP - OB = \frac{1}{OB} - OB = \frac{1 - OB^2}{OB} = \frac{AB^2}{OB} \quad \text{(recall (2))},$$

from which

$$BM = \frac{1}{2} BP = \frac{AB^2}{2OB} \tag{6}$$

Now,

$$OM = OB + BM = OB + \frac{AB^2}{2OB},$$

and then (3) yields

$$
\begin{aligned}
MC^2 = OM^2 - 1 &= \left(OB + \frac{AB^2}{2OB}\right)^2 - 1 \\
&= \left(\frac{2OB^2 + AB^2}{2OB}\right)^2 - 1, \qquad \text{(recalling (2))} \\
&= \left(\frac{1 + OB^2}{2OB}\right)^2 - 1 = \frac{1 + 2OB^2 + OB^4}{4OB^2} - 1 \\
&= \frac{1 - 2OB^2 + OB^4}{4OB^2} \\
&= \left(\frac{1 - OB^2}{2OB}\right)^2 = \left(\frac{AB^2}{2OB}\right)^2 = BM^2 \quad \text{from (6) above,}
\end{aligned}
$$

as desired.

12. Let the required masses at vertices A, B, C be x, y, z, respectively (Figure 8). Since the triangle is acute angled, the altitudes lie completely inside the triangle. In order to have the center of mass G at the orthocenter, the center of mass of the masses at B and C must occur at the foot D of the altitude from A. Hence we need

$$yc \cos B = zb \cos C;$$

similarly, we also require

$$ya \cos B = xb \cos A \quad \text{and} \quad xc \cos A = za \cos C.$$

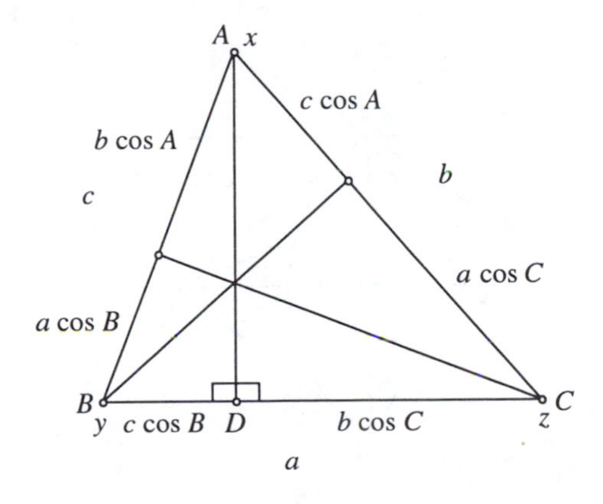

FIGURE 8

Thus

$$x = a \cos B \cos C$$
$$y = b \cos C \cos A$$

and

$$z = c \cos A \cos B$$

are suitable values, and all proportional values.

13. Each a_i is of the form $2^m 3^n, m, n \geq 0$, and the reciprocals of all such integers is contained in the product of the two infinite series

$$P = \left(1 + \frac{1}{2} + \frac{1}{2^2} + \cdots\right)\left(1 + \frac{1}{3} + \frac{1}{3^2} + \cdots\right).$$

Hence

$$\frac{1}{a_1} + \frac{1}{a^2} + \cdots + \frac{1}{a_n}$$
$$< P = \frac{1}{1 - \frac{1}{2}} \cdot \frac{1}{1 - \frac{1}{3}} = 2 \cdot \frac{3}{2} = 3.$$

14. Let CA meet BP at Q (Figure 9). Then, since O is the midpoint of BC and OP is parallel to CAQ (both are perpendicular to AB), P is the midpoint of BQ. Now, median CP bisects every line across $\triangle BCQ$ which is parallel to BQ, in particular AD (both AD and BQ are perpendicular to BC).

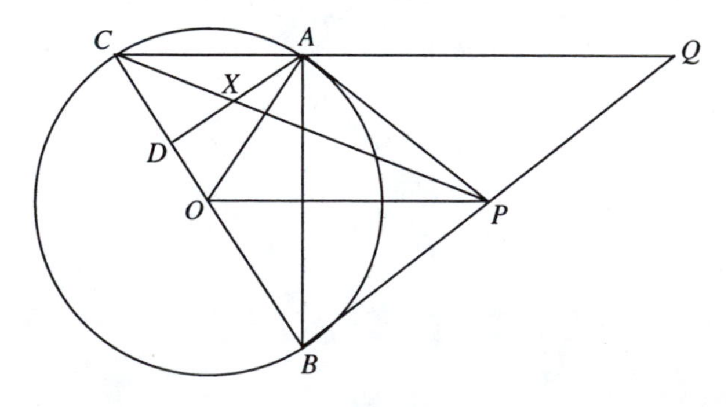

FIGURE 9

15. If $2\sqrt[3]{2} = x$ and $5\sqrt[3]{5} = y$, then $x^3 = 16$, $y^3 = 625$ and $x^3 - y^3 = (x-y)(x^2+xy+y^2) = 16 - 625 = -609$, a rational number.

    Hence

$$\begin{aligned}\frac{1}{2\sqrt[3]{2}-5\sqrt[3]{5}} &= \frac{1}{x-y} = \frac{1}{x-y}\cdot\frac{x^2+xy+y^2}{x^2+xy+y^2}\\ &= \frac{x^2+xy+y^2}{x^3-y^3} = \frac{x^2+xy+y^2}{-609}\\ &= -\frac{4\sqrt[3]{4}+10\sqrt[3]{10}+25\sqrt[3]{25}}{609}.\end{aligned}$$

16. Recall that the straight line L through the midpoints S and T of two parallel chords AB and CD is a diameter of a parabola and is parallel to the axis (Figure 10). A chord GH that is perpendicular to L is therefore perpendicular to the axis, and since the axis bisects every chord that is perpendicular to it, the line perpendicular to GH through its midpoint M is in fact the axis, and its intersection with the parabola is the vertex V. Also, the perpendicular VW to the axis at the vertex is the tangent at the vertex.

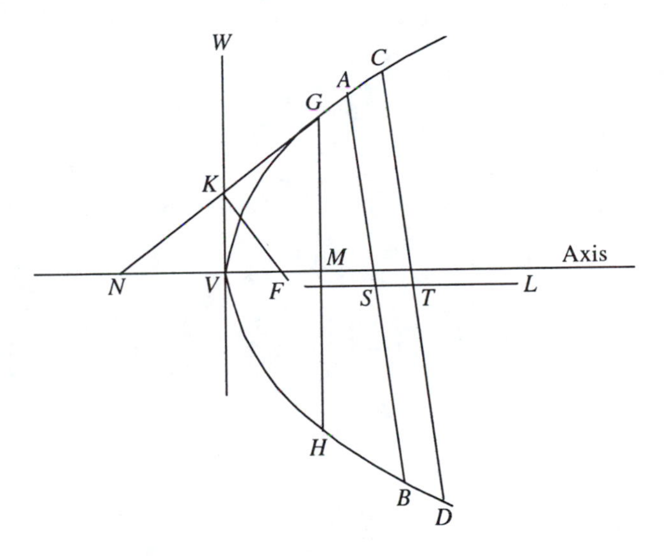

FIGURE 10

Now, marking off $VN = VM$ on the extended axis makes NG a tangent to the parabola, and every tangent to a parabola crosses the tangent at the vertex (VW) at the point (K), which is the foot of the perpendicular to the tangent from the focus. Thus the perpendicular to NG at K goes through the focus F, which is therefore the point where this perpendicular crosses the axis.

17. Since $t_n \cdot t_{n+1} = n$, we have $t_{n+1} = \frac{n}{t_n}$, and from any point in the sequence, the terms proceed

$$t_n, \quad n \cdot \frac{1}{t_n}, \quad \frac{n+1}{n} \cdot t_n, \quad \frac{n(n+2)}{n+1} \cdot \frac{1}{t_n}, \quad \frac{(n+1)(n+3)}{n(n+2)} \cdot t_n,$$
$$\frac{n(n+2)(n+4)}{(n+1)(n+3)} \cdot \frac{1}{t_n}, \quad \frac{(n+1)(n+3)(n+5)}{n(n+2)(n+4)} \cdot t_n, \quad \ldots.$$

Hence

$$\{t_{\text{odd}}\} = \{t_1, t_3, t_5, \ldots\}$$
$$= \left\{t_1, \frac{2}{1} \cdot t_1, \frac{2 \cdot 4}{1 \cdot 3} \cdot t_1, \frac{2 \cdot 4 \cdot 6}{1 \cdot 3 \cdot 5} \cdot t_1, \ldots.\right\}$$

tending to the limit

$$L = \frac{2 \cdot 4 \cdot 6 \cdot \cdots}{1 \cdot 3 \cdot 5 \cdot \cdots} \cdot t_1.$$

Now, Wallis' formula for π is

$$\frac{\pi}{2} = \frac{2 \cdot 2 \cdot 4 \cdot 4 \cdot 6 \cdot 6 \cdot \cdots}{1 \cdot 3 \cdot 3 \cdot 5 \cdot 5 \cdot 7 \cdot \cdots},$$

and so

$$L = \sqrt{\frac{\pi}{2}} \cdot t_1.$$

Similarly,

$$t_{\text{even}} = \{t_2, t_4, t_6, \ldots\} = \left\{\frac{1}{t_1}, \frac{1 \cdot 3}{2} \cdot \frac{1}{t_1}, \frac{1 \cdot 3 \cdot 5}{2 \cdot 4} \cdot \frac{1}{t_1}, \ldots\right\},$$

with

$$\lim t_{\text{even}} = \frac{1}{L}.$$

But, since

$$\lim \frac{t_n}{t_{n+1}} = \frac{\lim t_n}{\lim t_{n+1}} = 1,$$

we have

$$\lim t_{\text{odd}} = \lim t_{\text{even}}.$$

Hence

$$L = \frac{1}{L},$$
$$L^2 = 1,$$
$$\frac{\pi}{2} \cdot t_1^2 = 1,$$

and

$$t_1 = \sqrt{\frac{2}{\pi}}.$$

18. Clearly

$$N = (7^7)^7 \cdot 7^7 \cdot 77 = 7^{57} \cdot 11,$$

and therefore

$$\begin{aligned}\log N = 57 \cdot \log 7 + \log 11 &= 57(.845098\ldots) + 1.041392\ldots\\ &= 48.170588\ldots + 1.041392\ldots\\ &= 49.21198\ldots,\end{aligned}$$

implying that N has exactly 50 digits. Hence either

(a) some digit occurs at least six times or

(b) each of the ten decimal digits occurs exactly five times.

In case (b), the sum of the digits in N would be

$$5(0 + 1 + 2 + \cdots + 9) = 5(45) = 9 \cdot 25,$$

a multiple of 9, implying N itself is divisible by 9. But this is not so, for

$$N = 7^{57} \cdot 11 \equiv (-2)^{57} \cdot 2 \equiv [(-2)^3]^{19} \cdot 2 \equiv 1^{19} \cdot 2 \equiv 2 (\text{mod}\, 9).$$

Hence case (a) must hold.

19. The number obtained by removing d and adding $2d$ is

$$k = \frac{n - d}{10} + 2d,$$

giving

$$10k = n + 19d.$$

Since 10 and 19 are relatively prime, then 19 divides n if and only if 19 divides k. Because this holds for each number produced by the process, it is valid for the final result r.

This test is not as obliging as the test for divisibility by 9, in which case a number and its "digital root" actually have the same remainder upon division by 9. For $n = 39$, for example, the final result of the present test is 21, which leaves a remainder of 2 when divided by 19, while 39 leaves a remainder of 1.

20. (Solution due to John Rigby, University College, Cardiff)

Let the numbers be $a_1, a_2, \ldots, a_n$. Now, the product

$$\begin{aligned} s \cdot \left(\frac{1}{a_1} + \frac{1}{a_2} + \cdots + \frac{1}{a_n}\right) & \\ &= (a_1 + a_2 + \cdots + a_n)\left(\frac{1}{a_1} + \frac{1}{a_2} + \cdots + \frac{1}{a_n}\right) \\ &= n + \sum_{i<k}\left(\frac{a_i}{a_k} + \frac{a_k}{a_i}\right), \end{aligned}$$

and since $\frac{a}{b} + \frac{b}{a} \geq 2$ for all positive real numbers a and b, it follows that

$$\begin{aligned} s \cdot \left(\frac{1}{a_1} + \frac{1}{a_2} + \cdots + \frac{1}{a_n}\right) &\geq n + 2\binom{n}{2} \\ &= n + 2 \cdot \frac{n(n-1)}{2} \\ &= n + n^2 - n \\ &= n^2. \end{aligned}$$

Hence

$$\left(\frac{1}{a_1} + \frac{1}{a_2} + \cdots + \frac{1}{a_n}\right) \geq \frac{n^2}{s},$$

from which the desired minimum value is $\frac{n^2}{s}$, for equality **is** attainable (when the a's are all equal, and only then).

21. Let the exterior angle at A be $2x$ (Figure 11); then

$$\angle MEB = \angle MAB = x = \angle NAC = \angle NFC.$$

Also,

$$\angle BME = \angle BAE = \angle FAC = \angle FNC \ (= \angle A),$$

and triangles MBE and NCF are equiangular and therefore similar.

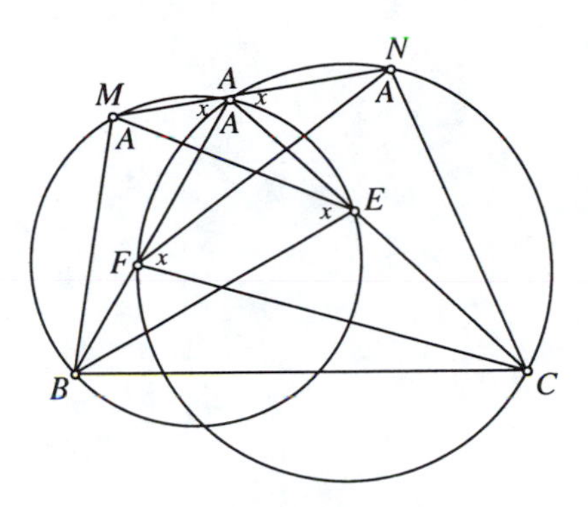

FIGURE 11

Now, in $\triangle MBE$, we have $\angle M = \angle A$ and $\angle E = x$, and since at vertex A we have $\angle A + 2x = 180°$, it follows that the third angle in the triangle,

$$\angle B = x \quad (= \angle E).$$

Hence $\triangle MBE$ is isosceles, and also the similar triangle $\triangle NCF$.

Subject Index

It is hoped that these brief descriptions will identify a topic sufficiently to direct you to its place in the book. The contents are listed under three headings.

1. Combinatorics and Combinatorial Geometry **Page**

2. Geometry **Page**

3. Algebra, Number Theory, Probability, and Calculus — Page